Life through Time

Evolutionary Activities for Grades 5–8

by

Kevin Beals, Nicole Parizeau, and **Rick MacPherson**
with **Kimi Hosoume** and **Lincoln Bergman**

Skills

Observing • Inferring • Using Evidence • Making Models • Comparing
Analyzing Data • Recording • Working Cooperatively • Communicating • Theorizing
Logical and Critical Thinking • Drawing Conclusions

Concepts

Paleontology • Evolutionary Biology • Change and Stasis • Adaptation
Form and Function • Respect for Organisms • Environmental Awareness
Continental Drift • Geologic Time • Chronology • Extinction

Themes

Evolution • Models • Systems and Interactions • Patterns of Change
Interdependence • Abundance and Diversity • Scale and Structure

Mathematics Strands

Number • Measurement • Pattern • Logic and Language
Data Interpretation • Drawing to Scale • Time Scale
Symmetry • Statistics and Probability

Nature of Science and Mathematics

Creativity and Constraints • Interdisciplinary Connections
Real-Life Applications

Time

Two class periods of 80–120 minutes and five class periods of 60–90 minutes

LHS GEMS®

Great Explorations in Math and Science
Lawrence Hall of Science
University of California at Berkeley

Lawrence Hall of Science,
University of California,
Berkeley, CA 94720-5200

Director: Ian Carmichael

Cover Design
Lisa Haderlie Baker

Internal Design
Lisa Klofkorn

Illustrations
Lisa Haderlie Baker
Lisa Klofkorn

**Aquarium and Terrarium
Background Illustrations**
Lisa Haderlie Baker

Director: Jacqueline Barber

Associate Director: Kimi Hosoume

Associate Director: Lincoln Bergman

Mathematics Curriculum Specialist:
Jaine Kopp

GEMS Network Director: Carolyn Willard

GEMS Workshop Coordinator:
Laura Tucker

Staff Development Specialists:
Lynn Barakos, Katharine Barrett, Kevin Beals,
Ellen Blinderman, Gigi Dornfest,
John Erickson, Stan Fukunaga, Linda Lipner,
Karen Ostlund

Distribution Coordinator: Karen Milligan

Workshop Administrator: Terry Cort

Trial Test & Materials Manager:
Cheryl Webb

Financial Assistant: Vivian Kinkead

Distribution Representative:
Fred Khorshidi

Shipping Assistant: Justin Holley

Director of Marketing & Promotion:
Steven Dunphy

Senior Writer/Managing Editor:
Nicole Parizeau

Editor: Florence Stone

Principal Publications Coordinator:
Kay Fairwell

Art Director: Lisa Haderlie Baker

Senior Artists: Carol Bevilacqua, Lisa
Klofkorn

Staff Assistants: Marcelo Alba, Haleah
Hoshino, Moe Moe Htwe, Kamand Keshavarz

Contributing Authors: Jacqueline Barber,
Katharine Barrett, Kevin Beals,
Lincoln Bergman, Susan Brady,
Beverly Braxton, Mary Connolly, Kevin Cuff,
Linda De Lucchi, Gigi Dornfest,
Jean C. Echols, John Erickson, David Glaser,
Philip Gonsalves, Jan M. Goodman,
Alan Gould, Catherine Halversen,
Kimi Hosoume, Susan Jagoda, Jaine Kopp,
Linda Lipner, Larry Malone,
Rick MacPherson, Stephen Pompea,
Nicole Parizeau, Cary I. Sneider, Craig Strang,
Debra Sutter, Herbert Thier,
Jennifer Meux White, Carolyn Willard

Initial support for the origination and publication of the GEMS series
was provided by the A.W. Mellon Foundation and the Carnegie Corpo-
ration of New York. Under a grant from the National Science Foundation, GEMS Leaders Workshops have been held across the country.
GEMS has also received support from: Employees Community Fund of
Boeing California and the Boeing Corporation; the people at Chevron
USA; the Crail-Johnson Foundation; the Hewlett Packard Company; the
William K. Holt Foundation; Join Hands, the Health and Safety Educational Alliance; the McDonnell-Douglas Foundation and the
McDonnell-Douglas Employee's Community Fund; the Microscopy Society of America (MSA); and the Shell Oil Company Foundation. GEMS
also gratefully acknowledges the contribution of word-processing equip-
ment from Apple Computer, Inc. This support does not imply responsibility for statements or views expressed in publications of the GEMS
program. For further information on GEMS leadership opportunities, or
to receive a catalog and the *GEMS Network News*, please contact GEMS.
We also welcome letters to the *GEMS Network News*.

International Standard Book Number: 0-924886-67-6

Printed on recycled paper with soy-based inks.

Library of Congress Cataloging-in-Publication Data

Beals, Kevin.
  Life through time : evolutionary activities for grades 5-8 / by
Kevin Beals, Nicole Parizeau, and Rick MacPherson ; with Kimi
Hosoume and Lincoln Bergman.
    p. cm.
  ISBN 0-924886-67-6 (alk. paper)
  1. Evolution (Biology)--Study and teaching (Elementary)--
Activity programs. 2. Evolution (Biology)--Study and teaching
(Middle school)--Activity programs. I. Parizeau, Nicole, 1957-
II. MacPherson, Rick. III. GEMS (Project) IV. Title.
QH362.B43 2002
372.3'5--dc21

                                2002151772

Wondering what's on the cover?
See pages 330–331 for a key to
the organisms.

ACKNOWLEDGMENTS

Life through Time, straightforward as we hope it is to use, was extraordinarily complex to produce! Its achievement involved many devoted and talented people here at LHS, for whose efforts we're very grateful.

With care, inventiveness, patience, and finesse, artists Lisa Baker and Lisa Klofkorn met—and surmounted!—the daunting task of representing life on Earth throughout history. This guide would be a pale shadow of itself without their beautiful work. Editor Florence Stone valiantly and expertly attended to the continuity and attention to detail a guide like this requires, provided excellent production suggestions, and kept us on geological track in the final stages. We're grateful to Carolyn Willard, a wonderful educator and colleague, who willingly plunged into the guide and gave astute advice that made it—particularly the final session—more intuitive and consistent for classroom presentation. On a complex trial-test unit such as this, we're reminded of the yeoman's job done by everyone who prepares our guides for local and national trial testing and assembles kits for all the classrooms involved. Over the course of Life through Time's evolution, members of that awesome team included Trial Test & Materials Manager Cheryl Webb and her predecessor Vivian Tong, and our invaluable student assistants Stacey Touson and Thania Sanchez.

We are extremely grateful for the supportive participation of Dr. Niles Eldredge, Chair of the Committee on Evolutionary Processes and Division of Paleontology (Invertebrates) at the American Museum of Natural History in New York. Dr. Eldredge reviewed parts of this guide in its penultimate stages, and we thank him for his promptness, enthusiasm, and generous endorsement.

Thank you to Walter Alvarez, Professor of Geology in the Earth and Planetary Sciences Department at the University of California, Berkeley for his generous time on the subject of Pangaea's formation and break-up. Thanks to Christopher Harris, who lent his hominid brain to help create the "Adapt or Die" camp activities that inspired this guide, and who willingly, enthusiastically, and convincingly played characters from liverwort to human. We're grateful to the UC Berkeley Botanical Gardens for lending us plants during early testing. Appreciation to Don Luce, one author's high school drama teacher, who encouraged him to read the play Inherit the Wind and later allowed himself to be persuaded to stage it.

Many enthusiastic and supportive teachers served as reviewers for Life through Time in its trial stages, and we deeply appreciate their valuable comments, classroom ideas, and samples of student work. This unit has been refined in large part thanks to their time and feedback. Please see page 320 for a complete list of reviewers' names. ■

In Memory
of
Stephen Jay Gould

As this GEMS guide approached completion, we joined
the world in mourning the loss of Stephen Jay Gould,
evolution's greatest champion since Darwin. His passion
for science, exuberant support of teaching, and sweeping
intellectual curiosity were unparalleled. His kindness,
support, and advice to one author remain key moments
in his life. We dedicate this guide to Steve Gould's
memory, and hope the activities inspire young minds to
further explorations in evolution and life through time.

Stephen Jay Gould
1941–2002

Contents

TIME FRAME

Life through Time is a particularly ticklish unit for which to suggest a time frame. Depending on the age and experience of your students, the length of your class periods, and your teaching style, the time needed for this unit may vary. The times given below are only suggestions; many teachers find that after they've warmed to the set-up routine in Session 2, the subsequent sessions take less time than estimated. Others may wish to extend the unit by combining it with other, related curricula (see "Pathways through the Unit" on page 6).

The most time will be needed for the first two sessions, as you acquaint students with the concepts of change over time and the "time-travel" stations. Don't be alarmed! After the initial set-up and overviews of the first time-travel session ("Early Life"), the next four will flow more smoothly. See "Advance Map of the Time-Travel Sessions" on page 9 for a bird's-eye view of the core sessions; this'll give you an idea of the "rhythm" of the guide.

You may choose to assign certain parts of the time-travel sessions (students' predictions for the next time period, for instance) as homework, to shorten class time. Session 7 may require more than one session to complete, and/or you may choose to have student groups finish their assignments after class. Try to build flexibility into your schedule so that you can expand the number of class sessions if necessary.

Session 1: The Tree of Life .... 80–120 minutes

Session 2: Early Life .... 80–120 minutes

Session 3: The Invasion of Land .... 60–90 minutes

Session 4: Fish and Amphibians .... 60–90 minutes

Session 5: Reptiles .... 60–90 minutes

Session 6: Birds and Mammals .... 60–90 minutes

Session 7: Reflecting on Life through Time .... 60–90 minutes

WHAT YOU NEED FOR THE WHOLE UNIT

The quantities below are based on a class size of 32 students. Depending on the number of students in our class, you may, of course, need different amounts of materials.

This list gives you a concise "shopping list" for the entire unit. Please refer to the "What You Need" and "Getting Ready" sections for each session. They contain more specific information about the materials needed for the class and for each team of students.

Many photocopies are necessary for this unit. You may want to enlist a volunteer to do the copying. Also, the copying could be done as needed for each session, or all the necessary pages could be copied at one time. Further, the copied pages could be organized by using a distinct color paper for each session.

Nonconsumables

- ❑ 1 length of adding-machine tape to represent the whole unit's **Class Time Line;** choose length from chart in "Getting Ready" of Session 1 (page 15)
- ❑ 8 copies of the **Students' Tree of Life Cards** (page 26)
- ❑ 1 copy of **Fossils—Teacher's Answer Sheets** (pages 64–68; 5 pages total)
- ❑ 1 copy of **"Guts"—Teacher's Answer Sheet** (page 73)
- ❑ 1 copy each of **Most Representative Organism Script** for Sessions 2–6 (pages 69–70, 117–118, 152–153, 186–187, and 218)
- ❑ 1 copy of **title sign** for **Life through Time** wall chart (page 76)
- ❑ 1 copy each of these **headers** for **Life through Time** wall chart (page 77):
  - ___ Ages
  - ___ Aquarium
  - ___ Terrarium
  - ___ "Guts"
  - ___ Continental Drift
- ❑ 1 copy each of **Tree of Life Organism Cards** for Sessions 2–5 (pages 79–81, 122–126, 158–159, 191–192) to add to the **Tree of Life** wall chart
- ❑ 1 copy each of **Major Evolutionary Events—Time Period #1–Time Period #5** (pages 82, 127, 160, 193, 223)

- ❏ 2 large cloths (such as sheets or large towels) to cover and conceal the aquarium and terrarium until you're ready to show them
- ❏ several containers to "house" organisms at the stations (Use whatever containers seem appropriate for the organism, such as an old food storage container for flatworms.)
- ❏ 2 or more natural bath sponges
- ❏ 1 water spray bottle, to keep earthworms moist

An overhead transparency of each of the following:
- ❏ **Tree of Life Branch—Arthropods** (page 23)
- ❏ **Sample Tree of Life Cards** (page 24)
- ❏ **From Worm to Insect** (page 25)
- ❏ **"Guts": Single-Celled Organism/Sponge/Jellyfish/ Earthworm** (page 72)
- ❏ **Early Life** (page 74)
- ❏ **Other Early Invertebrates** (page 75)
- ❏ **Moss Reproduction** (page 119)
- ❏ **Early Land Plants** (page 120)
- ❏ **Arthropods** (page 121)
- ❏ **Conifer Reproduction** (page 154)
- ❏ **Fish** (page 155)
- ❏ **Amphibians** (page 156)
- ❏ **Conifers** (page 157)
- ❏ **Flowering Plant Reproduction** (page 188)
- ❏ **Reptiles** (page 189)
- ❏ **Flowering Plants** (page 190)
- ❏ **Birds** (page 219)
- ❏ **Mammals** (page 220)
- ❏ **Skeletons through Time** (page 221)
- ❏ **"Guts": Bird and Mammal** (page 222)
- ❏ *(optional/recommended)* **Algae Reproduction** (page 71)

Two copies each of the following station sheets:
- ❏ **Early Life** (page 74)
- ❏ **Other Early Invertebrates** (page 75)
- ❏ **"Guts": Single-Celled Organism/Sponge/Jellyfish/ Earthworm** (page 72)
- ❏ **Fossils—Time Period #1–Time Period #5** (pages 83, 128, 161, 194, 224)
- ❏ **Algae Reproduction** (page 71)
- ❏ **Early Land Plants** (page 120)

❑ **Arthropods** (page 121)
❑ **Fish** (page 155)
❑ **Amphibians** (page 156)
❑ **Conifers** (page 157)
❑ **Reptiles** (page 189)
❑ **Flowering Plants** (page 190)
❑ **Birds** (page 219)
❑ **Mammals** (page 220)

Three copies each of the following station sheets:
❑ **Continental Drift—Time Period #1–Time Period #5;** two each for the stations and one each for the **Life through Time** wall chart (pages 84, 129, 162, 195, 225)
❑ **"Guts": Bird and Mammal** (page 222)

For the Time Travel Aquarium:
❑ 1 aquarium tank (Any size can work, but the larger the tank, the easier it will be to fit increasing numbers of plastic and real animals and plants as the unit progresses. We recommend no smaller than the classic five-and-a-half-gallon size. See the chart of tank dimensions on page 63.)
❑ aquarium gravel and/or sand (Coarse gardening sand, sandbox sand, or sand-blasting sand will all work fine. **Note: do not use coral sand or coral gravel;** these are only used for saltwater aquariums and can kill freshwater organisms. Do not use sand or gravel you have collected from an ocean beach, as it will be salty. If your sand or gravel is dusty, rinse it thoroughly before use.)
❑ 1 copy each of **Aquarium Background—Time Period #1– Time Period #5** (pages 92A, 136A, 160A, 200A, 224A) *See the chart of tank dimensions on page 63 for helpful information about sizing the background to your tank size.*
❑ 1 natural (not synthetic) bath sponge (You'll need more for a separate station; see earlier in Nonconsumables.)
❑ 1 (or more) plastic jellyfish (To make your own, see sidebar on page 39.)
❑ freshwater snail(s) (Can be purchased in aquarium stores or found in freshwater ponds and streams.)
❑ aquatic insects (Can be caught with a net in a local pond or stream.)
❑ sea urchin shell(s)
❑ plastic sea star(s)

- ❑ plastic mollusk(s), or real shells such as clam or mussel shells
- ❑ artificial coral (do not collect coral from the wild)
- ❑ plastic sharks
- ❑ plastic bony fishes
- ❑ goldfish, mosquito fish, or whatever kind of live fish you have (Small fish can be caught in a local pond with a net. We do *not* recommend putting your student's pet fish in the Time Travel Aquarium. Please see note on page 21 about mosquito fish.)
- ❑ plastic crocodiles, turtles, and plastic prehistoric marine reptiles— pleisosaurs, ichthyosaurs, mososaurs, etc.
- ❑ plastic whales and other plastic marine mammals
- ❑ plastic sharks, including a large plastic great white shark (*Carcharodon carcharias*) model if you have one (to represent the giant *Carcharodon megalodon,* related to the present-day great white)

For the Time Travel Terrarium:
- ❑ 1 aquarium tank (Again, the larger the tank, the more diorama elements can be added. See the chart of tank dimensions on page 63.)
- ❑ enough lava dirt (preferable) or lava rock to make a 2- or 3-inch-deep layer in the terrarium (If you cannot find these, you can use other rocks. **Soil is a last resort—there *was* no soil on Earth in the period covered in Session 2.**)
- ❑ 1 copy each of **Terrarium Background—Time Period #1– Time Period #5** (pages 92B, 136B, 160B, 200B, 224B) *See page 63 for help in sizing the background to your tank size.*
- ❑ enough soil to make a 2- or 3-inch-deep layer in the terrarium
- ❑ at least one of the following early plants:
    - __ moss (Can be obtained from yards, or purchased in bags from a nursery or hardware store.)
    - __ liverwort (Can be obtained from a nursery or garden; it's an herb with broad heart-shaped leaves.)
    - __ club moss (Ground pine or princess pine are examples of club mosses living today.)
    - __ horsetail
    - __ small ferns
    - __ ginkgo (Popular urban ornamental tree in people's yards or on West Coast sidewalks.)
    - __ cycad (Early plant resembling—but different from!—a palm or tree-fern.)

___ cones and needles, or just cones, from any type of cone-bearing tree (pine, redwood, fir, etc.) Ideally, get both male cones (they're the small ones) and female cones (larger) from a single type of tree.

❑ plastic scorpion(s)

❑ plastic millipede(s)

❑ live land snail(s) (*Note:* If the land snails are kept in an open container, they may escape and eat nearby paper products…such as unlaminated background illustrations—yikes! They'll also eat your plants.)

❑ plastic dragonfly

❑ plastic salamander

❑ any plastic dinosaurs and early terrestrial reptiles, including lizards, snakes, and tortoises

❑ one or two small flowering plants (such as sweet alyssum or any other easily available)

❑ beetles, ants, hissing cockroaches, or other real or plastic insects

❑ acorns

❑ plastic butterfly

❑ plastic or plush opossum

❑ any plastic early mammals (wooly mammoth, saber-toothed cat, etc.)

❑ any plastic modern mammals, birds, and other animals (but not domesticated animals such as dogs)

❑ more small flowering plants

❑ grasses

For the volcano options:

❑ 2–3 glass or plastic chemistry vials or tubes, plus one extra to use for adding vinegar during the activity (see "Resources," page 307)

❑ 2–3 plastic or foam cups, each with a hole punched in its bottom, to place over the vials to create the slopes of the volcanoes

❑ enough brown clay for sticking pieces of volcanic rock to the rim of the vials or the sides of the cups, for a more realistic appearance

❑ 1 or more small red bicycle lights with a steady shining option, to make the volcanoes "glow" from within

❑ same number of sealable baggies, to protect the bicycle light(s) from rubble and moisture

Some or all of the following organisms for the stations:

- ❏ algae (Can be obtained from a local birdbath or pond; "scum" or film on the inside of a fish tank or other body of fresh water. May look fuzzy, or like green hair. Seaweed can also be used.)
- ❏ 5 or so earthworms (Can be obtained from the ground or compost pile. Can also be purchased at bait shops or biological supply houses; see "Resources.")
- ❏ 2 or more flatworms (Also known as *Planaria*. Can sometimes be collected in freshwater streams and ponds, if you know what you're looking for, or purchased from some aquarium stores or biological supply houses. See "Resources.")
- ❏ a handful of tubifex worms (Also known as "blood worms." Can be purchased from local aquarium store or biological supply houses; see "Resources.")
- ❏ *Triops* (tadpole shrimp) eggs, hatched into adults (It takes a few days to a week before they're large enough to observe. See "Making It as Easy as Possible" on page 8, and "Resources.")
- ❏ isopods—pill bugs (roly-polies) and sow bugs
- ❏ hermit crab★
- ❏ crayfish ("crawdad")
- ❏ freshwater snails
- ❏ centipede
- ❏ millipede
- ❏ spider
- ❏ clam
- ❏ sea star★
- ❏ silverfish
- ❏ moss
- ❏ liverwort
- ❏ club moss (ground pine or princess pine are living examples of early land plants)
- ❏ invertebrate exoskeleton (shed!)
- ❏ trilobite fossil
- ❏ fish (goldfish, tetra, guppies, etc.)
- ❏ horsetail
- ❏ fern
- ❏ salamander★
- ❏ frog★
- ❏ cockroach
- ❏ fish skeleton
- ❏ dead fish, whole or partly dissected
- ❏ amphibian skeleton

- ❏ lizard
- ❏ snake
- ❏ turtle
- ❏ tortoise
- ❏ insects—especially those with complete, four-stage metamorphosis, such as ants, butterflies, and beetles
- ❏ flowering plants
- ❏ large flower(s) cuttings
- ❏ magnolia-tree flower or stem with leaves
- ❏ acorns
- ❏ cypress tree branch
- ❏ hickory nut
- ❏ reptile egg (do not collect viable eggs from the wild)
- ❏ shed reptile skin
- ❏ reptile skull or skeleton
- ❏ any safe mammal
- ❏ any safe bird
- ❏ bird skull
- ❏ mammal skull
- ❏ grasses

★These organisms in particular are delicate creatures and require careful handling and special care. In many states (including California) it is ILLEGAL to remove organisms from the wild.

Additional materials for Option 1 in Session 7:
- ❏ 1 overhead transparency of **Standard Geologic Time Line** (page 253)
- ❏ 1 overhead transparency of **Time Line Comparison Chart** (page 254)
- ❏ 8 sets of **Period Information Sheets** (pages 255–268; 14 pages total per team)

Additional materials for Option 2 in Session 7:
- ❏ *(optional/recommended)* 1 overhead transparency of **Standard Geologic Time Line** (page 253)
- ❏ *(optional/recommended)* 1 overhead transparency of **Time Line Comparison Chart** (page 254)

Additional materials for Option 3 in Session 7:

- ❑ 1 copy of **Explaining Major Evolutionary Change** (page 269)
- ❑ *(optional/recommended)* 1 overhead transparency of **Standard Geologic Time Line** (page 253)
- ❑ *(optional/recommended)* 1 overhead transparency of **Time Line Comparison Chart** (page 254)

Additional materials for Option 4 in Session 7:

- ❑ 1 overhead transparency of **Standard Geologic Time Line** (page 253)
- ❑ 1 overhead transparency of **Time Line Comparison Chart** (page 254)
- ❑ 1 paper copy **Time Line Comparison Chart** (page 254) to hang in the classroom
- ❑ 11–16 copies of the **Period Information Sheet** for the period teams of students will be working on (pages 255–268)

Consumables

- ❑ 32 copies of the **Organism Key** (pages 29–34; 6 pages total)
- ❑ 32 copies of the *Life through Time* **Scavenger List for Parents** (pages 21–22), if needed
- ❑ 5 copies of **The Age of** _____ sign (page 78); one each for Sessions 2–6 (*Note:* You may choose instead just to write this information on a blank 8 ½" x 11" sheet of paper— which would allow you to re-use the **Life through Time** wall chart for future classes—or directly on the chart after the election, to simplify.)
- ❑ 1 hard-shelled egg—a chicken egg will do (We recommend hard-boiled!)

For the Time Travel Journal:

- ❑ 32 **Time Travel Journal** cover pages (page 27)
- ❑ 32 journal pages labeled **First Life on Earth** (page 28)
- ❑ 32 sets each of **Time Travel Journal** pages labeled **Time Period #1, Time Period #2, Time Period #3, Time Period #4,** and **Time Period #5** (pages 85–93, 130–136, 163–170, 196–202, 226–232; 38 pages total) Don't copy these until you've added the session's organisms to the **Organism Adaptations** page; see "Getting Ready" for each session.

For the Time Travel Aquarium:
- ❑ enough dechlorinated water to fill and periodically top off the aquarium (You may use bottled spring water, but **do not use distilled water,** because it lacks beneficial minerals. You may also use tap water with dechlorinating liquid, which can be purchased at an aquarium store.)

For the Time Travel Terrarium volcano options:
- ❑ enough baking soda to fill two or three vials two-thirds full
- ❑ enough chocolate powder to add 1 tablespoon to each volcano eruption mixture, for better appearance
- ❑ enough red food coloring to add to the volcano eruption mix, for better appearance
- ❑ enough vinegar to occasionally add a few drops to each vial
- ❑ enough water to fill vials two-thirds full
- ❑ a few small pieces of dry ice to add to the water in the vials (see "Resources")

Additional materials for Option 4 in Session 7:
- ❑ *(optional)* 11–16 shoeboxes or other small containers (plastic, if any of the dioramas will be aquatic!) to be used for the dioramas (Students can bring these containers in; see note on page 208- in Session 6.)

General Supplies

- ❑ an overhead projector and screen (or other white background)
- ❑ 1 meterstick
- ❑ 1 globe or world map
- ❑ 8 pairs of scissors
- ❑ 8 envelopes
- ❑ 1 sheet of butcher paper 7 ft. wide x 6 ft. tall for **Life through Time** wall chart
- ❑ 1 sheet of butcher paper 3 ft. wide x 5 ft. tall for **Tree of Life** wall chart
- ❑ paper towels, to provide cover for the earthworms at station
- ❑ either a large binder clip or a three-ring binder, for the **Time Travel Journal**
- ❑ chalkboard, chart paper, or a blank overhead transparency for listing questions and reminders for the class (see each "Getting Ready" in Session 7)

❑ 32 pieces of blank or lined paper on which to write the assignment in Session 7

❑ *(optional)* laminator, if you wish to protect and reuse the wall chart signs and/or tank illustrations

❑ *(optional)* three-hole punch, if you wish the students to keep their **Time Travel Journals** in a three-ring binder

❑ *(optional)* mural paper and colored pens and pencils

❑ *(optional)* chart paper for making signs

Additional materials for Option 4 in Session 7:

❑ additional materials with which to make the dioramas: poster paper, scissors, reference books, dirt, plants, plastic animals and plants from the Time Travel Aquarium and Terrarium

INTRODUCTION

"The theory of evolution is quite rightly called the greatest unifying theory in biology." —*Ernst Mayr, Professor of Zoology, Emeritus, Harvard University*

If only we could travel into the past, to see what the world was like and piece together how life evolved! In a way, we do: this is the work of paleontologists. They're not time travelers, but they do gather all the evidence they can to recreate the distant past. Details of our picture of the past are constantly changing, with new discoveries of fossils and more accurate dating methods, but the basic framework has been in place for many years.

How the Unit Works

Because evolution is such a huge concept, it cannot be taught in its entirety in one unit. In this guide we've chosen to focus on **changes and stability in life through time** as an introduction to biological evolution. This is a huge subject in itself, and the guide does not, for instance, delve in detail into *how* evolution works, or explore genetics.

Over the course of this guide, your students will observe changes in life structures and behaviors from the simple to the complex over large chunks of time. Students will begin to fit together some foundational ideas, such as the relationship between form and function in organisms, to such larger, overarching concepts as life's expansion into a terrestrial environment from an aquatic one. They'll see how geologic time relates to the evolution of life through time.

The activities in *Life through Time* present the abstract world of Earth's past as concretely as possible, as students watch the past "come to life" before their eyes. This unit combines a variety of topics popular with students:

- observing living animals and plants
- examining fossils and bones
- creating dioramas with a variety of materials

A central strategy of this guide is creation of a **Time Travel Aquarium** and a **Time Travel Terrarium,** set up by the teacher and/or students. Using plastic model animals, illustrations, live plants, and a few live animals, you and your class will set up dioramas for each of five time-travel periods. These become two of several stations through which the students rotate in Sessions 2 through 6. The dioramas are fascinating to

Chronicling evolution is really a matter of capturing an "evolutionary snapshot"; what paleontologists understand at a given moment. This knowledge expands continually, as technologies grow and more sites on Earth become accessible for research. When we began this guide, the earliest evidence of hominids dated to approximately three million years ago. In the last few months of its production, Sahelanthropus tchadensis (nicknamed "Toumaë") was discovered in Chad, Central Africa—placing our earliest hominid ancestors on Earth six or seven million years ago. No doubt evolutionary events will continue to be repositioned as our understanding grows.

students, and really allow them to visualize what scientists think the past was like. A variety of materials can be used to achieve these understandings—but of course the more living organisms you use, the more intriguing it will be for your students.

Students get to observe life change through time in a variety of ways. They study the aquarium and terrarium each day and observe how the flora and fauna have changed, or "evolved," from the previous session. They add major evolutionary events to a **Class Time Line** posted around the room. They watch as **Life through Time** and **Tree of Life** wall charts build each day as they add more illustrations depicting

Life through Time
and
National Content Standards

This guide focuses on the big changes in the history of life on Earth and supports life science content objectives in the *National Science Education Standards*. In particular, the 5th–8th grade standard under "Diversity and Adaptations of Organisms" directly addresses the concepts developed in *Life through Time*:

- Millions of species of animals, plants, and microorganisms are alive today. Although different species might look dissimilar, the unity among organisms becomes apparent from an analysis of internal structures, the similarity of their chemical processes, and the evidence of common ancestry.

- Biological evolution accounts for the diversity of species developed through gradual processes over many generations. Species acquire many of their unique characteristics through biological adaptation, which involves the selection of naturally occurring variations in populations. Biological adaptations include changes in structures, behaviors, or physiology that enhance survival and reproductive success in a particular environment.

- Extinction of a species occurs when the environment changes and the adaptive characteristics of a species are insufficient to allow its survival. Fossils indicate that many organisms that lived long ago are extinct. Extinction of species is common; most of the species that have lived on Earth no longer exist.

animal and plant life and continental drift. They observe animals and plants representing the time period, and study their adaptations. In creative assignments toward the end of the unit, the students draw on their accumulated knowledge to express their ideas in writing or oral presentation.

Is This a Unit about Evolution?

This is a guide about how life has evolved over time. Evolution is the single most powerful concept in the life sciences. It's the big, unifying idea that ties the complexities of life together. Although evolution can be a controversial topic in some communities, trying to teach life sciences while ignoring evolution would be like teaching students about marine animals while imagining that the ocean doesn't exist, or teaching chemistry without the periodic table of elements. From the scientific perspective on life, evolution is everywhere. It explains how things came to be the way they are.

Choosing Geologic Time Spans

In looking at geologic time, we've consciously chosen NOT to focus on the names of established periods, eras, and epochs, but rather on the big and important concepts throughout life's evolution. In Sessions 2–6, students study each big-picture time period and then **name the "age" themselves,** choosing what they perceive to be the most representative organism of the time. This serves to reinforce the big ideas and concepts, which are most important, and helps students become invested in and retain the information. Not until Session 7 are students introduced to the classic geologic time line (eras, periods, epochs, etc.), so that they can make connections between the geologic record and their own observations of life through time.

For your own reference, standard geologic periods—Precambrian Time, Mesozoic Era, and so forth—are provided as marginal notes in "Summary of Sessions," below; in every time-travel session overview; and in "Background for the Teacher" on page 270.

Summary of Sessions

Session 1: The Tree of Life
Students sort pictures of organisms from the past and present, and construct a partial branch of the "tree of life." They're also introduced to the time-travel simulation idea that works as a framework for the unit.

The concept of evolution has an importance in education that goes beyond its power as a scientific explanation. All of us live in a world where the pace of change is accelerating. Today's children will face more new experiences and different conditions than their parents or teachers have ever had to face in their lives.

...To accept the probability of change—and to see change as an agent of opportunity rather than as a threat—is a silent message and challenge in the lesson of evolution.
—National Academy of Sciences Teaching about Evolution and the Nature of Science

The diversity of life on Earth is the outcome of evolution: an unpredictable and natural process of temporal descent with genetic modification that is affected by natural selection, chance, historical contingencies and changing environments.
—National Association of Biology Teachers (NABT)

Finally, they post the **Class Time Line** to which they'll add major evolutionary events as the unit progresses.

Session 2: Early Life

This is the first time-travel session of the unit. Students rotate through stations representing geologic time from 4 billion, 500 million years ago to 544 million years ago: 4,500–544 MILLION YEARS AGO, or MYA. (Archeologists and other scientists commonly use the abbreviations BYA and MYA to mean "BILLION YEARS AGO" and "MILLION YEARS AGO.")

Precambrian
Time

The focus for this session is on **sponges, jellyfish, worms, and single-celled organisms, with a recommended optional station on algae reproduction.** The stations for this session include:

- the Time Travel Aquarium and Terrarium
- live organisms representing those of the time
- picture sheets with questions (provided in the guide)

Students record observations in their **Time Travel Journals.** At the end of the session, the teacher leads a wrap-up discussion and places the day's accumulated information on the **Life through Time** wall chart, the day's major evolutionary events on the **Class Time Line,** and the day's organisms on the **Tree of Life** wall chart.

This session, like the next four, concludes with a comic skit of "stump" speeches by organisms vying for the title of "most representative organism of the time." Students then elect the organism they think best represents the age they studied.

Session 3: The Invasion of Land

Cambrian,
Ordovician,
and Silurian
Periods

Similar stations, recording, wrap-up, and election as in Session 2, with the focus on **early land plants and invertebrates** such as sea squirts, trilobites, sea scorpions, and shellfish. The session's time period covers 544–410 MYA.

Session 4: Fish and Amphibians

Devonian and
Carboniferous
Periods

Similar stations, recording, wrap-up, and election, with the focus on **fish, amphibians, and cone-bearing plants.** The session's time period covers 410–286 MYA.

Session 5: Reptiles

Similar stations, recording, wrap-up, and election, with the focus on **reptiles and early flowering plants.** The session's time period covers 286–65 MYA.

Permian, Triassic, Jurassic, and Cretaceous Periods

Session 6: Birds and Mammals

Similar stations, recording, wrap-up, and election, with the focus on **birds, mammals, and more complex flowering plants,** including grasses. The session's time period covers 65 MYA–present time.

Cenozoic Era, which includes the Tertiary and Quaternary Periods

Session 7: Reflecting on Life through Time

In one of four creative options, this wrap-up session brings together the knowledge students have acquired over the course of the unit.

- **Option 1: Time Traveler Adventure Stories.** After class discussion on the standard geologic time line, small student groups use **Period Information Sheets** to write fictional but scientifically accurate stories of adventures in the past.

- **Option 2: Dramatizing Life through Time.** Students produce songs, skits, poems, or other representations of life's evolution over time. Brief class discussion on the standard geological time line is recommended, but optional.

- **Option 3: Explosions and Extinctions.** Students come up with evidence and explanations for the extinction, decline, or population explosion of one or more species. Can be a writing assignment, mural with oral description, or skit. Brief class discussion on the standard geological time line is recommended, but optional.

- **Option 4: 3-D Class Diorama.** After class discussion on the standard geologic time line, student teams use **Period Information Sheets** to create scenes of specific geologic time periods. They conclude by putting them all together to form a three-dimensional classroom model of life through time.

Pathways through the Unit

This unit includes jumping-off points for just about all aspects of life science, allowing many creative options for presentation. One way of teaching this unit is to present it on seven to 10 consecutive days; this allows you to gather materials all at once. An alternative is to present Sessions 1 & 2 and 6 & 7 consecutively, but allow time between the time-travel sessions: 2, 3, 4, 5, and 6. This allows your class to spend more time understanding the changes and concepts of each time-span. (It also allows you more time to gather materials.)

If you choose the second option, you can space these sessions a day (or a week) apart, interspersing activities from other curricula appropriate to the time period. After pollination is introduced in Session 5, for example, you can do a more in-depth study of pollination. When bacteria are introduced in Session 2, other activities involving bacteria can be inserted. An insect unit can be presented after insects are introduced in Session 3.

In fact, this unit can be the "spine" of a school year's life science curriculum, with each of the time-travel sessions (2, 3, 4, 5, and 6) separated by a month (which would provide lots of setup and material-gathering time).

Important Evolutionary Themes

This unit is extremely rich with content, and essential information is provided in "Background for the Teacher" beginning on page 270. Several main ideas thread through the unit and should be emphasized over the course of each session; these are described on pages 289–292 of the background. (You may wish to start there.)

Important Notes on Preparation, Materials, and Scheduling

The most exciting element of this unit is its materials; they're also the most challenging aspect. The materials are extremely exciting for the students to handle but can be challenging for you, the teacher, to prepare. But the more you put into the materials, the more your students will enjoy and learn from their "traveling through time" experience! It's important to note that setup will be most difficult and time-consuming for Session 2; for all sessions after that, setup is more routine.

Before beginning the unit, be sure to allow ample time—a month or more—to check sources for organisms and materials, and to plan how and when the organisms will arrive in the classroom. Whatever your schedule, this has to be taken into account in your planning. This preparation can be quite time-consuming the first time you teach this unit. Doing things at the last minute can result in extra expense and problematic delays. The more you can obtain items through donation and other alternative methods and sources, the lower the costs and the higher your confidence will be. Once you have what you need, the preparation time for each activity is quite reasonable. (See "Resources" on page 307.)

We strongly recommend that you read the guide through before beginning and enlist help in the advance preparation. As in many other activity-based science units, the preparation load can be lessened considerably if you collaborate with one or more other teachers. Parents, grandparents, or other adult volunteers can also provide much assistance. (See also "Making It as Easy as Possible," below.)

We emphasize these considerations *not* to discourage you in any way from presenting the activities, but to make sure you're aware of them from the start. The many teachers who tested these activities strongly affirmed that the educational impact of this unit was well worth it. The plants and animals chosen for these activities are quite accessible and the expenses reasonable, given the high return in long-term learning!

As always, we welcome your ideas and suggestions. If you find good sources for organisms or materials used in *Life through Time,* write a letter to the *GEMS Network News,* call, or send us an e-mail (our address is on the inside front cover). GEMS guides are revised frequently based on continuing teacher feedback, and we welcome your comments!

There are four categories of materials you can use to re-create the past. **We suggest using a combination of all four categories:**

- **Live animals and plants are best,** and should be included whenever possible. We can't actually travel to the past and see the real organisms, so each session includes live animals and plants that are *similar* to those from the time being studied. Many can be found in yards, gardens, or other outdoor areas. Others need to be purchased at a store, or mail-ordered. **(Please see the important note on collecting and handling living**

organisms on page 21.) The unit can be taught without live plants and animals—but keep in mind, your students will greatly benefit from observing living things up close!

- **Non-living but real natural-history specimens are second best.** The unit provides an opportunity to pull out any biological specimens, skulls, bird eggs, reptile eggs, shed skins, or whatever may be available to you in your school and use them in a meaningful way. A local high school, college, or nature center may have specimens for you to borrow.

- **Plastic animals and plants are third best.** Many students of this age have outgrown playing with plastic toy dinosaurs but are still intrigued by them and enjoy an excuse to use them in a way that is not "childish." Building dioramas with three-dimensional models will help them visualize life in the past.

- **Fourth best are illustrations.** Sources include biology and zoology textbooks, resource guides, magazines, and natural-history or science journals.

Making It as Easy as Possible

- **Draft your students to help assemble materials.** We've included an advance list to send home to families (pages 21-22). You may want to mail these out well ahead of presenting the unit, or you can send them home with students at the end of Session 1. Many of the materials in this unit can be acquired in this way. Students enjoy bringing in plastic toys or their small pets, and there's usually a group in each class that enjoys capturing roly-polies and worms. The list is fairly skeletal; you'll probably choose to add more information before sending it out, such as the date by which (or on which) you need, say, a particular plant, or if you'd like all materials on the same date. If you're seeking tanks of a specific size for the Time Travel Aquarium and Terrarium, be sure to include the dimensions in your request.

- **Be sure to give yourself enough time to collect and order** the organisms and other materials you'll be using. Session 5, for example, uses live *Triops* (tadpole shrimp, *Triops longicaudatus*), tiny crustaceans that have survived since the

Triassic! *Triops* can be purchased at many science-related toy stores or ordered from a supply house (see "Resources" on page 307), and **should be raised a few days to a week before Session 5 is presented** (they're initially too small to see). Remember, you're trying to set *scenes* for your students that are as stimulating and "real" as possible. The extra effort to collect the necessary materials will be repaid many times over when you see your students make the leap to those difficult and abstract concepts!

- **Rely on your students** to help with station preparation and maintenance. For example, a great after-school/extra-credit project is to have students put together the aquarium and terrarium dioramas for the upcoming session. These need to change in most of the sessions; a time-consuming effort for you, but a definite active-learning project for a group of interested students!

- **Recruit fellow teachers** to present the unit along with you. This definitely helps with scheduling and materials-preparation issues, and gives you a partner with whom to work out bigger concerns of guiding student learning. You may choose to team-teach the unit or to teach it separately but share material setups. Either way, it's a big help in presenting a complex but exciting unit.

- **Anticipate the space.** Ideally, your classroom can accommodate one aquarium and one terrarium that remain set up as dioramas throughout the unit. Some wall space is needed to hang the **Tree of Life** wall chart (3 feet wide x 5 feet tall), the **Life through Time** wall chart (7 feet wide x 6 feet tall), and the **Class Time Line** (choose the length your wall space allows, from page 15). If you can't keep the dioramas or charts set up between classes, don't worry; the charts can be rolled up and set aside until needed, and the dioramas can be moved out or dismantled and reassembled by your team of student helpers.

Advance Map of the Time-Travel Sessions

Here's a peek at how the middle five sessions of *Life through Time* are structured. We hope you'll be reassured to see that these time-travel sessions are essentially presented the same way each time; only some of

Please note, as we say on page 42, that much of the preparation for Session 2 (the first time-travel session) will contribute to the overall set-up of the whole unit. Also, the exciting and elaborate (and optional!) volcano set-up in Session 2 occurs only once.

the components change. (Please see "Summary of Sessions" on page 3 and "Teacher's Outline" on page 294 for further details on all sessions of the unit.) All the time-travel sessions break down pretty much the same way:

Getting Ready, Sessions 2 through 6:

1. **Copy and cut up the Most Representative Organism Script** into parts the students will play.
2. **Fill in Organism Adaptations page** for the students' **Time Travel Journals.**
3. **Copy and make sets of the journal pages** for the new time period.
4. **Copy the overhead transparencies** for the session.
5. **Copy and cut up the Tree of Life Organism Cards.**
6. **Update the Time Travel Aquarium and Time Travel Terrarium** with organisms for the new time period. (In Sessions 3 and 6, you'll need to replace the old with the new; most of the time, you can just keep adding organisms.)
7. **Copy the Major Evolutionary Events sheet** for the new time period.
8. **Update the stations** for the new time period, replacing station signs on continental drift and fossils, and replacing organism–adaptation specimens. Some sessions have special stations ("guts" or "plant reproduction," for example), which need to be set up.

Moving through Sessions 2 through 6:

1. **Distribute the students' Time Travel Journals** and **Organism Keys.**
2. **Distribute new journal pages** for the students to insert.
3. **Conduct a brief station overview.** This goes faster after the first time, as several of the stations have exactly the same assignments from session to session (just the elements are new).
4. **Begin the station rotation.** (Allow 45–60 minutes for this.)
5. **Conduct a station debrief.** Discuss each of the stations using focus ideas and material in the guide. You may be using overheads, charts, or brainstorms.
6. **Update the Tree of Life wall chart** with pictures of the session's organisms (the background illustrations from the aquarium and terrarium).
7. **Update the Class Time Line** with the session's **Major Evolutionary Events** sheet.
8. **Conduct the campaign and election** (i.e., have students act out and vote) for the **Most Representative Organism of the Age.**

9. **Post the strip of adding-machine tape** representing the next session's time period on the **Class Time Line** (except in Session 6, the last time-travel session).

If You're Teaching the Unit to Multiple Classes

If you'll be presenting this unit to more than one class, here are a couple of options for the time-travel sessions:

1. Conduct the station debriefs as written, remembering to take down and reposition the wall chart illustrations, major evolution-ary events, and tank backgrounds between classes.

or

2. Stop after the station rotations in each session, BEFORE the debriefs. (We've marked these "pausing spots" in the unit like this: **P**) This allows you to keep the stations assembled, the background illustrations taped to the aquarium and terrarium, and the wall charts pristine for a subsequent class. Once all the classes have finished the station rotations, you can pick up where you left off: Add the organism, station, and tank illustrations to the wall charts, update the time line, and explain that these come from the earlier setup.

These are just options; if you have enough time and/or are teaching the unit to just one class, you'll probably want to run straight through each session. ■

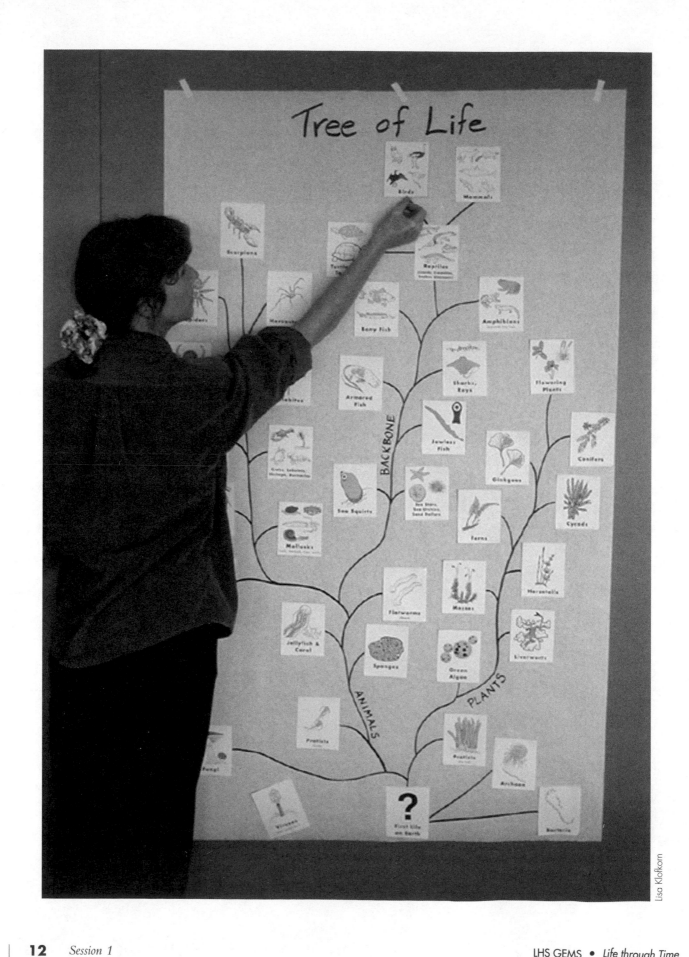

Overview

Life through Time begins with a discussion of what life was like long ago. To get them thinking about the subject, and to help you assess what they already know, students write down their ideas and share them with the class. Later, this can help both you and your students gain a sense of how much more they've learned over the course of the unit.

In this session, students identify the features scientists use to group related organisms. Students also begin to hypothesize how these organisms may have evolved through time. Working in small groups, teams sequence drawings of past and present organisms in the order they think represents simplest to most complex. They use these to construct a branch of the "tree of life," illustrating how living things have changed from early organisms to modern-day life-forms.

Next, students are introduced to the central idea behind the unit: "time travel" back to five time periods in Earth's history. They'll be observing Earth and its life-forms in each period while concentrating on the changes that happen over huge stretches of time. **Time Travel Journals** and an **Organism Key** are distributed and explained as ways of keeping track of what students observe, discover, and learn throughout the unit.

■ What You Need

For the class:

- ❑ 1 overhead transparency of **Tree of Life Branch—Arthropods** (page 23)
- ❑ 1 overhead transparency of **Sample Tree of Life Cards** (page 24)
- ❑ 1 overhead transparency of **From Worm to Insect** (page 25)
- ❑ 1 length of adding-machine tape to represent the whole unit's **Class Time Line;** choose your desired length from the chart in "Getting Ready" (page 15)
- ❑ an overhead projector and screen (or other white background)
- ❑ 1 meterstick
- ❑ 1 globe or world map

For each team of four students:
- ❏ 1 copy of the **Students' Tree of Life Cards** (page 26)
- ❏ 1 pair of scissors
- ❏ 1 envelope

For each student:
- ❏ 1 **Time Travel Journal** cover page (page 27)
- ❏ 1 journal page labeled **First Life on Earth** (page 28)
- ❏ 1 copy of the **Organism Key** (pages 29–34; 6 pages total)
- ❏ 1 copy of the *Life through Time* **Scavenger List for Parents** (pages 21–22), if needed

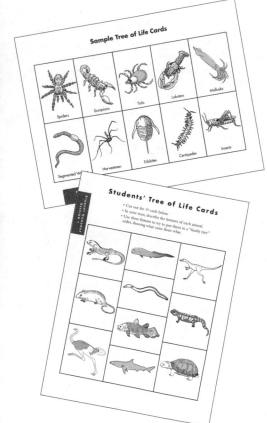

■ Getting Ready

1. Make one copy of each transparency. Cut up the **Sample Tree of Life Cards** transparency (page 24) into cards that can be manipulated on the overhead projector. (**Do NOT** cut up the **Tree of Life Branch—Arthropods** transparency.)

2. Make one copy of **Students' Tree of Life Cards** (page 26) for each team of students. **Do not cut the sheets.**

3. Make one copy of the **Time Travel Journal** cover page (page 27) and **First Life on Earth** journal page (page 28) for each student.

4. If needed, customize and copy one *Life through Time* **Scavenger List for Parents** (pages 21–22) for each student.

5. Copy and staple one **Organism Key** (pages 29–34; 6 pages total) for each student.

6. Have scissors and envelopes ready to give to each team of students.

7. From the chart on the next page decide which overall length of adding-machine tape best fits your classroom for the **Class Time Line.** The longer a time line you choose, the more "major evolutionary events" you'll be able to record for each period.

8. On the chart, look up the overall length you chose for your **Class Time Line.** Mark, then cut up, the adding-machine tape into the

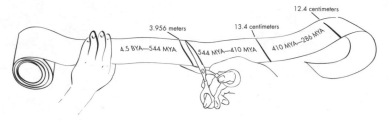

lengths designated in the appropriate column. Label each piece with its span of time. **You'll be adding one of these strips to lengthen the Class Time Line toward the end of every time-travel session.**

Class Time Line Lengths			
	Choices of Overall Length		
	4 METERS 50 CM	9 METERS	13 METERS 50 CM
Session 2 4.5 BYA-544 MYA This period lasted 3,956 million years.	3.956 meters (3m 95cm 6mm)	7.912 meters (7m 91cm 2mm)	11.868 meters (11m 86cm 8mm)
Session 3 544 MYA-410 MYA This period lasted 134 million years.	13.4 cm (13cm 4mm)	26.8 cm (26cm 8mm)	40.2 cm (40cm 2mm)
Session 4 410 MYA-286 MYA This period lasted 124 million years.	12.4 cm (12cm 4mm)	24.8 cm (24cm 8mm)	37.2 cm (37cm 2mm)
Session 5 286 MYA-65 MYA This period lasted 221 million years.	22.1 cm (22cm 1mm)	44.2 cm (44cm 2mm)	66.3 cm (66cm 3mm)
Session 6 65 MYA-present time This period lasted 65 million years.	6.5 cm (6cm 5mm)	13 cm	19.5 cm (19cm 5mm)

This chart is a mathematical representation of all time on Earth. The overall length of 4.5 meters represents the 4.5 billion years of Earth's existence, and every session in the column below is a portion of that total. For instance, Session 2 represents the time period from 4.5 BYA to 544 MYA; that's 3,956 million years—or a 3.956 meter portion of the total 4.5 meters of all time.

*The overall lengths of 9 meters and 13.5 meters in the chart are simply multiples of the 4.5-meter length. They allow the **Class Time Line** to be longer, to accommodate more "major evolutionary events."*

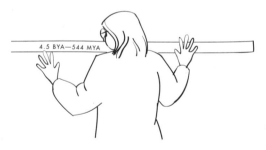

4.5 BYA—544 MYA

9. Post the **Class Time Line** strip for Session 2, "4.5 BYA–544 MYA," on the wall of your classroom. You'll refer to it at the end of Session 1. This is the first of five consecutive strips you'll post, so be sure to position it so the time line can "grow" horizontally.

10. Using a pencil, pen, or felt marker, draw a mark on your meterstick at the lengths representing one million, 10 million, 100 million, and one billion years, referring to the chart at right.

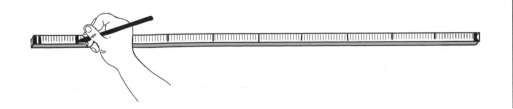

Marking Geologic Intervals on the Meterstick	
1 million years	1 mm
10 million years	1 cm
100 million years	10 cm
1 billion years	1 meter

■ **Introducing the Tree of Life**

1. Announce to the students that they're about to study what life was like many millions of years ago. In a sense, they'll be time travelers, going back in time to learn about the animals, plants, and environment of long ago, and studying how life has changed over large chunks of time. Tell students that scientists have found evidence that the first life on Earth appeared 3 billion, 500 million years ago!

2. Hand out the journal page **First Life on Earth** and have students write their names on them.

3. Ask students to jot down what they think the animals, plants, land, water, atmosphere, weather, and continents may have been like when life first appeared on Earth. Assure them that this is just to get them thinking, and all you need are their best guesses.

4. After a few minutes, ask a few students to share their ideas about early life with the class. If it's not mentioned, tell them scientists think the first life-form on Earth was a **single-celled organism.** Write the following definitions on the board and keep them posted throughout the unit:

   Sample Tree of Life Cards

   Spiders | Scorpions | Ticks | Lobsters | Mollusks
   Segmented Worms | Harvestmen | Trilobites | Centipedes | Insects

   OVERHEAD
   SESSION 1

   • **An organism is a living plant or animal.**
   • **A cell is the smallest structural living unit that can function independently.**

5. Place the cut-up **Sample Tree of Life Cards** transparencies on the overhead projector and briefly name them. Explain that the organisms range from very "simple" to "complex." In this case, "complex" means having different body parts doing different jobs (like a wing that flaps and a foot that walks), rather than all or most body parts doing the same job (like all segments in an earthworm acting the same way to move the worm along).

   Add this definition to the board:

   • **A single-celled organism is composed of one cell. It's the simplest life-form.**

6. Ask students which organism on the cards they think is the **simplest**—the least complex. [The segmented worm (earthworm).] Ask what features of the earthworm give them a clue that it's a simple organism. [Soft body composed of undifferentiated segments; no apparent sense organs; no apparent means of moving—no legs; all body parts functioning in a similar way.]

7. Ask which organism on the cards the students think is the **most complex.** [Probably the insect.]

8. Ask the class what kinds of clues we can get from living things to tell whether one organism is related to (descended from) another. [We can observe external features (features we can see), such as number of legs, body shape, type of teeth, etc., and also internal processes, such as digestion.]

9. Ask the students which organisms on the cards they think might have descended from (are related to) which others, based on simple and more complex features. Place the transparency cards in ascending order (most simple at the bottom to most complex at the top) according to the students' guesses. **It's very important to explain that the organisms at the "top" are NOT there because they're the "most advanced" or "best"; they're merely the most complex.**

10. Explain that hundreds of thousands of scientists, through decades and decades of research, have concluded that species change over time, or **evolve.** These changes are evident in their **adaptations**—characteristics that proved most successful for a species in its environment over many generations. Add these definitions to the board:

    • **An adaptation is a feature or behavior that can improve an organism's chance for survival.**
    • **Species change, or evolve, over many generations.**

11. Show the transparency **From Worm to Insect,** an artist's depiction of how segmented worms may have evolved into insects over millions of years. Through discussion, have students observe that hairs on segments of worms may have evolved into legs on segments of centipedes, which may have eventually evolved into organisms with claws, antennae, and mouth parts, having fewer legs.

12. Show the **Tree of Life Branch—Arthropods** transparency. Referring back to the "from worm to insect" discussion, tell students that after many years of study this diagram represents how scientists think these organisms (all in Phylum Arthropoda) are related to (descended from) one another.

13. Divide students into teams of four. Each team should have a clear desk or table on which to work.

It's a common misconception to think that life evolves in a long, linear march toward greater and greater complexity. Species have acquired more complex features and behaviors over time—but sometimes they've simplified, too. And complex life hasn't replaced all simple life. To this day, the most successful organisms on Earth are its absolute simplest life-forms: bacteria. (The book Full House, by Stephen Jay Gould, is a good teacher's read on this subject. See "Resources" on page 307.)

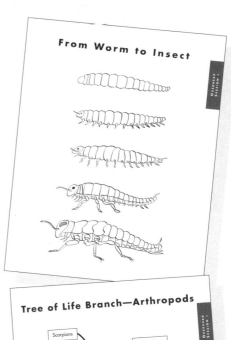

From Worm to Insect

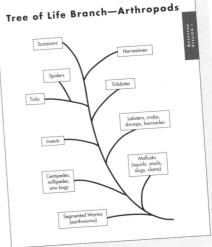

Tree of Life Branch—Arthropods

Evolutionary trees can be constructed in many ways. See pages 286–287 of "Background for the Teacher" for some historic examples.

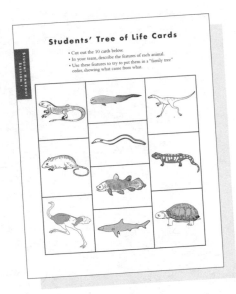

14. Tell students that their teams will go through the same steps you just went through as a class, using a different set of "Tree of Life" cards at their desks. Explain the procedure as follows:

    a. Each team will receive a sheet with 10 pictures of animals and plants; some from modern times and some from the past.

    b. Teams will cut apart the cards and discuss which organisms they think descended from which others. Remind them that they should begin by describing **external features** that provide clues about how simple or complex each organism is.

    c. Teams should then place the organisms in ascending order (from most simple at the bottom to most complex at the top) on the desk.

15. Hand out to each team the sheet called **Students' Tree of Life Cards,** a pair of scissors, and an envelope. When all teams are equipped, let them begin. Allow 15–20 minutes, depending on interest level and ability.

16. Call an end to the activity and ask for the students' attention. Explain that although scientists agree about many parts of the "tree of life," there are still other parts about which they don't agree. Tell the class that in the same way that scientists' trees of life may differ, their team's results may look different from another's. Have the teams circulate to view the work of others and see how they may have organized the pictures differently.

17. Ask a few teams to share their "trees of life" with the class and discuss the reasoning behind their organization of the organisms.

18. Have one student from each team collect the team's cut-up **Students' Tree of Life Cards,** put them in the envelope, and take the envelope and scissors to a designated location.

■ Introducing Time Travel

1. Tell students that for the rest of this unit they're going to do what many scientists probably wish they could do. They'll be time-traveling scientists! They'll visit **five different time periods** from Earth's past and study the animal and plant life in each of those times.

2. Tell them that of course they won't really be traveling through time, but they'll be observing one aquarium and one terrarium model

(diorama) of life in a different time for the next five sessions, representing each time period in turn. Explain to students that since most species of the past are now extinct, they'll be represented by:

- similar live animals and plants from today
- plastic animals and plants that represent the past
- illustrations of animals and plants from the past

3. If you didn't send scavenger lists to parents ahead of time, tell the students you'll be sending them home with a list for the family, to help collect items for the time-travel sessions.

4. Explain that what students will see is based on what hundreds of thousands of scientists have learned and hypothesized through centuries of work. Tell students they'll get a chance to watch the "tree of life" evolve!

5. Show the globe or world map. Ask students if they know how old Earth is. [It's approximately 4.5 billion (4,500,000,000) years old!] That means the planet had already been around for more than a billion years before the first organisms appeared.

6. Ask your students how they think scientists know about ancient organisms if these organisms aren't around today. [From fossils of bones and plants, and casts (molds) of animal tracks, and by dating fossilized rocks to determine their age.] Explain that as new fossils are found, our picture of the past becomes more accurate and thorough. A single new fossil find can provide evidence that something lived millions of years earlier than we'd thought!

7. Tell students that the first time period they'll be visiting includes the very earliest time for which we have evidence of life. This period is from four billion, five hundred million years ago (4.5 BILLION YEARS AGO, or 4.5 BYA) to five hundred forty-four million years ago (544 MILLION YEARS AGO, or 544 MYA). Introduce your students to the MYA and BYA abbreviations that many scientists use. You may wish to write out these numbers with all their zeros on the board, to help students grasp the huge amount of time; that's 4,500,000,000–544,000,000 years ago!

■ Introducing the Class Time Line

1. Point out the adding-machine-tape **Class Time Line** strip you cut for the next session, "4.5 BYA–544 MYA," posted on the wall.

4.5 BYA—544 MYA

How old is Earth?

No rocks on Earth can be dated back 4.5 billion years, because early in its history the Earth's surface was scalding hot, and the original rocks melted and were later re-formed into other rocks. So how do we know the Earth is older than that? See "Background for the Teacher" on page 270 for more.

If your students need more experience with large numbers, you might have them visit "Ask Dr. Math" online at http:// mathforum.org/dr.math/faq/ faq.large.numbers.html. See also "Resources" on page 307 for books about large numbers.

---

2. Hold your pre-marked meterstick up to the time line on the wall to show students how much each of these chunks of time (one million, 10 million, 100 million, and one billion years) takes up on the time line.

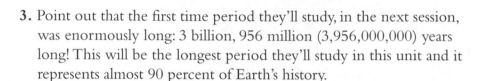

4.5 BYA—544 MYA

3. Point out that the first time period they'll study, in the next session, was enormously long: 3 billion, 956 million (3,956,000,000) years long! This will be the longest period they'll study in this unit and it represents almost 90 percent of Earth's history.

4. Distribute a **Time Travel Journal** cover and an **Organism Key** to each student. Have them write their names on the journal cover and page one of the Organism Key. Allow a few minutes for students to browse through the Organism Key and take in the numbers and variety of representative life-forms it depicts. Let them know they'll be using this key in the next several sessions.

5. Be sure students have written their names on their **First Life on Earth** journal pages. Collect these, the students' **Time Travel Journal** covers, and their **Organism Keys** for later use. Tell them their time travel will begin with the next session!

6. If appropriate, customize and distribute the *Life through Time* Scavenger **List for Parents** for each student to take home.

Life through Time Scavenger List for Parents

Dear Parent: Your child is about to participate in an exciting unit called *Life through Time*. Students will observe changes in life structures and behaviors from the simple to the complex over large periods of time. To accomplish this, we'll be assembling dioramas in the classroom, using a variety of items, to help students visualize life in the past.

WANTED:

Some dead, some alive.

———— ◆◆◆ ————

Can you help make our class *Life through Time* unit "come to life"? Here are some items we're looking for. Note: living organisms should be in appropriate containers, and items on loan should be labeled with the student's name.

Important Note on Collecting and Handling Living Organisms

All living creatures deserve our respect, our care, and a gentle touch. Whether your child is collecting earthworms from the compost bin or goldfish from the family aquarium, it's critical to model thoughtfulness and respect in the capture and handling of these creatures. Some organisms should definitely NOT be collected from the wild. Many—some butterflies and other insects; ocean organisms such as starfish; and several lizards and amphibians—are protected **by law.** (In some areas this goes for antlers, shells, carcasses, and bones, too.) Even non-threatened species may become stressed during capture or in captivity. Please use the utmost care and judgment when helping your child collect living organisms for this activity (especially the ones marked ★ in this list), and **be sure all creatures are carefully returned to their original locations as soon as the activity or unit is finished.** Your care and concern will provide a powerful role model for your children and help them become good stewards of our living resources.

Living plants:

- cones from any type of cone-bearing tree (pine, redwood, fir, etc.)
- small flowering plants (cuttings or with roots)
- acorns
- large flower cuttings from local garden, especially magnolia
- small ferns
- algae
- moss
- liverworts
- ginkgoes
- cycads (palm-like plants)
- grasses
- club moss (ground pine or princess pine are examples living today)
- horsetail rushes

liverworts

club moss

Living animals:

- earthworms
- land snails
- freshwater snails
- tubifex worms
- flatworms
- isopods—pill bugs (roly-polies) or sow bugs
- hermit crabs★
- crayfish (crawdad)
- centipedes
- millipedes
- spiders (do not handle spiders!)★
- clams
- silverfish (an insect)
- fish (goldfish, mosquito fish, or other) *NOTE: mosquito fish (Gambusia) can be an invasive species, competing with native aquatic life-forms when added to lakes and natural waterways.* **Be sure to dispose of mosquito fish into contained, artificial ponds or water gardens only.**
- salamanders★
- frogs★
- cockroaches
- triops (tadpole shrimp)

flatworms

centipedes

frogs

- lizards★
- snakes (not poisonous!)★
- turtles★
- tortoises★
- insects
- pet mammals (hamsters, gerbils, mice, rats, rabbits, etc.)
- pet birds (parakeets, parrots, canaries, etc.—not chickens or other fully domesticated species)

lizards

insects

Real, but not living, animals or animal remnants:

- sponges (real, not synthetic)
- sea urchin shells
- trilobite fossils
- corals (do not collect living corals from the wild!)
- hatched-out or blown-out reptile eggs
- hatched-out or blown-out emu or ostrich eggs
- shed reptile skins
- bird skulls
- mammal skulls
- shed antlers

trilobite fossils

mammal skulls

Plastic animals:

- plastic dinosaurs
- any plastic early mammals (wooly mammoth, saber-toothed cat, etc.)
- any plastic animals from today (except fully domesticated animals such as chickens)
- plastic sharks
- plastic jellyfish
- plastic scorpions
- plastic millipedes
- plastic sea stars
- plastic shellfish
- plastic dragonflies
- plastic sharks
- plastic bony fish
- plastic butterflies
- plastic crocodiles

plastic dinosaurs

plastic dragonflies

plastic bony fish

Other:

- 1 tank for aquarium and 1 tank for terrarium (preferred size: _____ gallons)
- lava dirt (preferable) or lava rock
- baking soda
- white vinegar
- 2 glass or plastic vials
- small red bicycle light, with a steady shining option
- brown clay
- chocolate powder
- red food coloring
- soil
- shoeboxes

Tree of Life Branch—Arthropods

OVERHEAD SESSION 1

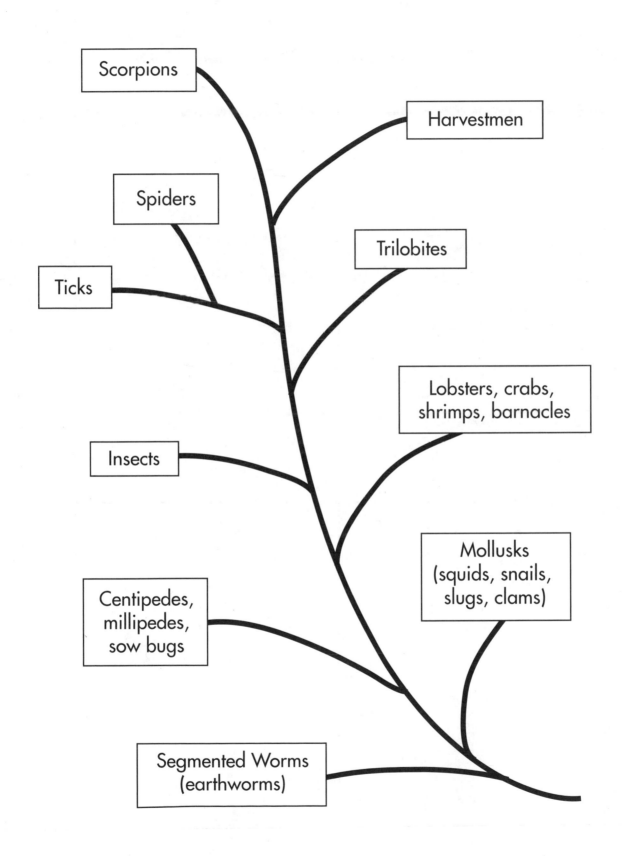

Scorpions

Harvestmen

Spiders

Trilobites

Ticks

Lobsters, crabs, shrimps, barnacles

Insects

Mollusks (squids, snails, slugs, clams)

Centipedes, millipedes, sow bugs

Segmented Worms (earthworms)

Sample Tree of Life Cards

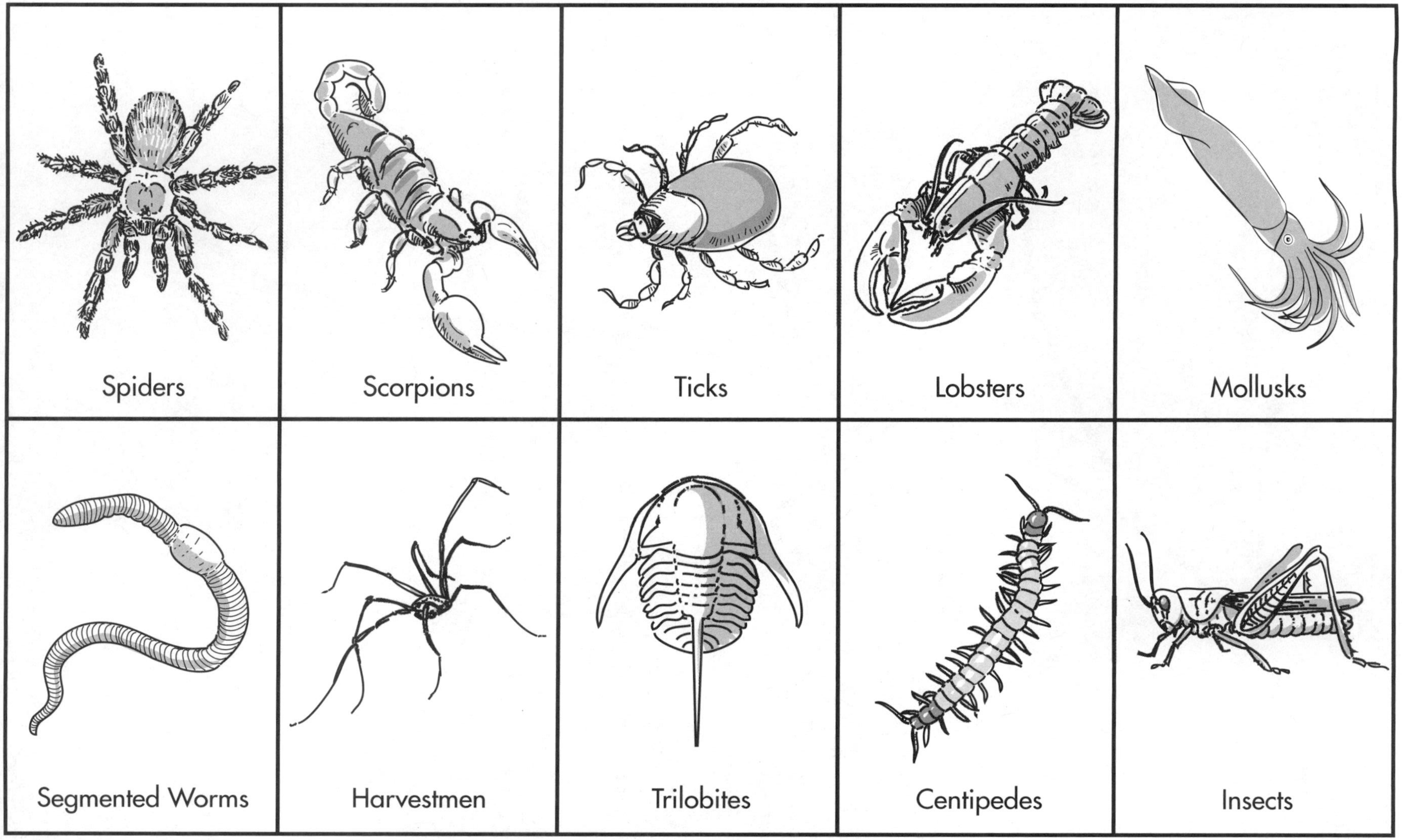

Segmented Worms

Spiders

Harvestmen

Scorpions

Trilobites

Ticks

Centipedes

Lobsters

Insects

Mollusks

From Worm to Insect

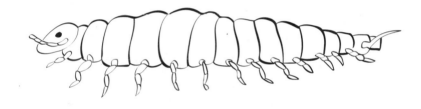

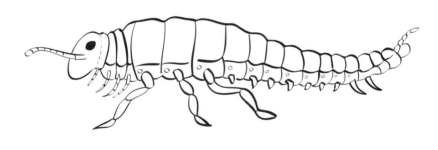

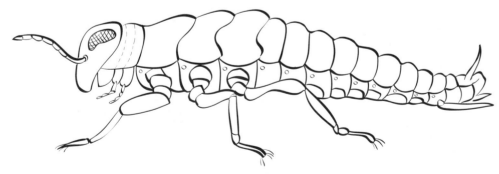

Students' Tree of Life Cards

- Cut out the 10 cards below.
- In your team, describe the features of each animal.
- Use these features to try to put them in a "family tree" order, showing what came from what.

TIME TRAVELS
A Journal for Time Travelers

LIFE THROUGH TIME

EVOLUTIONARY ACTIVITIES FOR GRADES 5-8

Name _____

Name _____

First Life on Earth

Write some of your ideas about what the very first life on Earth may have been like.

_____

_____

_____

_____

_____

_____

_____

_____

_____

_____

_____

_____

_____

_____

_____

_____

_____

_____

_____

Name _____

ORGANISM KEY

Single-Celled Organisms

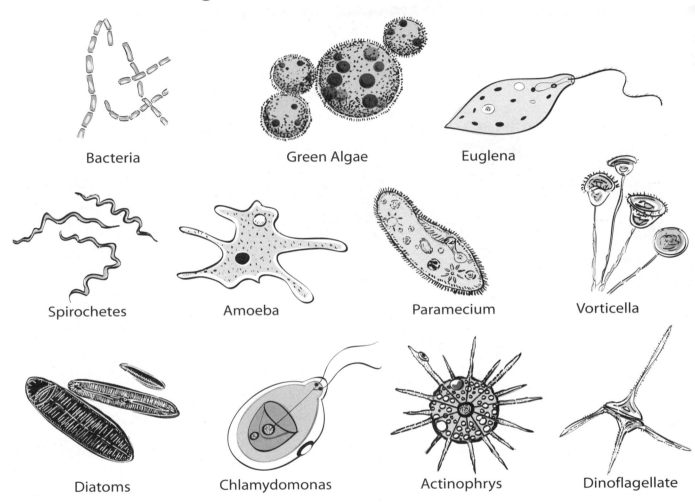

Bacteria   Green Algae   Euglena

Spirochetes   Amoeba   Paramecium   Vorticella

Diatoms   Chlamydomonas   Actinophrys   Dinoflagellate

Fungi

Bread Mold   Mushroom

Invertebrates

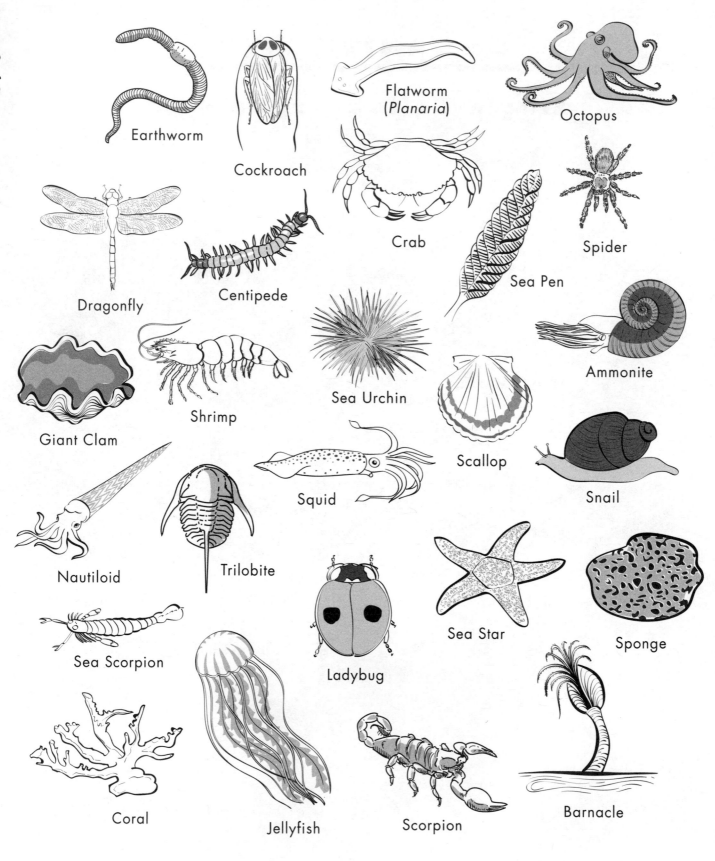

Earthworm

Cockroach

Flatworm (*Planaria*)

Octopus

Dragonfly

Centipede

Crab

Sea Pen

Spider

Giant Clam

Shrimp

Sea Urchin

Scallop

Ammonite

Nautiloid

Trilobite

Squid

Snail

Sea Scorpion

Ladybug

Sea Star

Sponge

Coral

Jellyfish

Scorpion

Barnacle

Plants

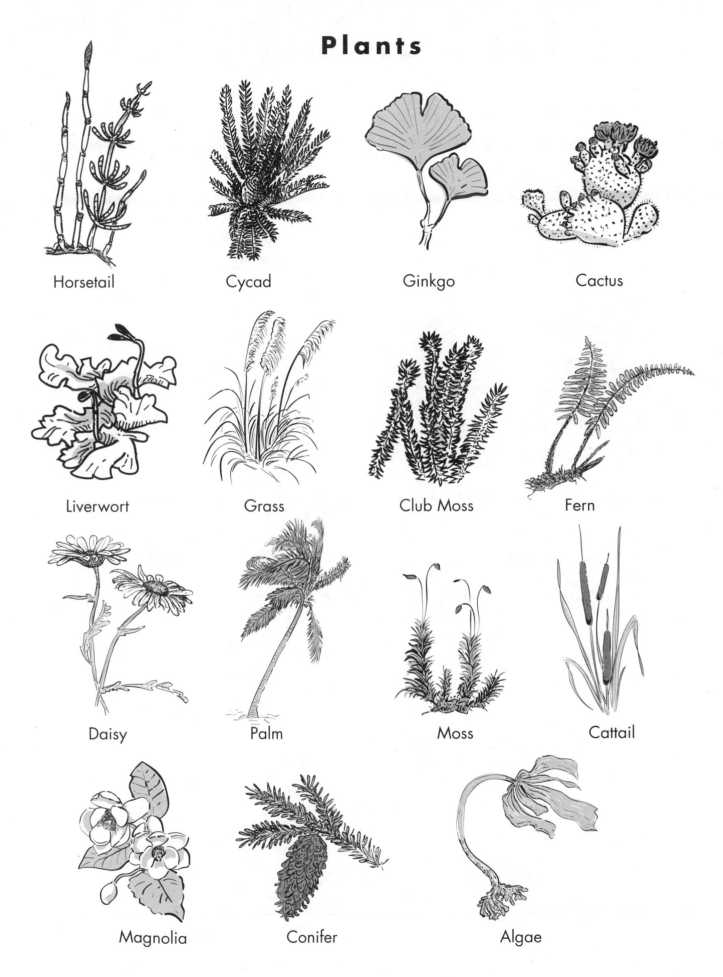

Horsetail

Cycad

Ginkgo

Cactus

Liverwort

Grass

Club Moss

Fern

Daisy

Palm

Moss

Cattail

Magnolia

Conifer

Algae

Amphibians and Reptiles

Plateosaurus

Tortoise

Diplocaulus

Eryops

Tyrannosaurus

Brachiosaurus

Deinosuchus

Salamander

Styracosaurus

Stegosaurus

Ankylosaur

Pterosaur

Triceratops

Camptosaurus

Discosauriscus

Snake

Coelurosaur

Iguanodon

Dolichosoma

Cynognathus

Crocodile

Turtle

Cetiosaurus

Acteosaurus

Frog

Apatosaurus

Lizard

Fish

Climatius

Birkenia

Whale Shark

Leopard Shark

Lungfish

Bluefin Tuna

Eurhinosaurus

Catfish

Ichthyosaurus

Perch

Barracuda

Pteraspis

Dunkleosteus

Eel

Seahorse

Hybodus

Birds

Owl

Ostrich

Cormorant

Archaeopteryx

Pelican

Vulture

Duck

Mammals

Giant Ground Sloth

Giraffe

Bull

Bat

Kangaroo

Wolf

Saber-toothed Cat

Rhinoceros

Bear

Coelodonta

Humpback Whale

Orca

Monkey

Gorilla

Seal

Elephant

Pakicetus

Horse

Platybelodon

Narwhal

Camel

Anteater

Woolly Mammoth

Hyracotherium

Tiger

Manatee

Rabbit

Gray Whale

Rytiodus

Raccoon

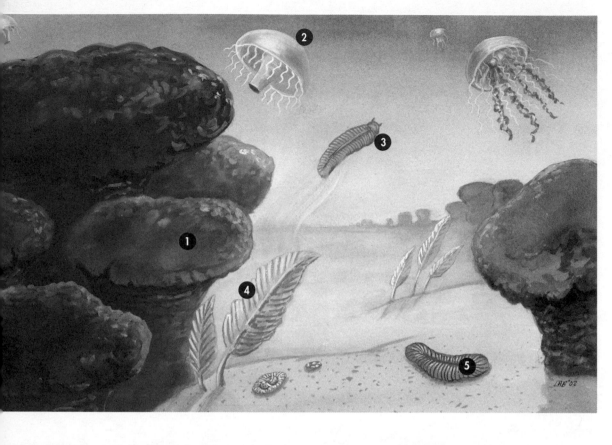

Aquarium

1. Stromatolites
2. Jellies
3. *Spriggina*
4. Sea Pens
5. *Dickinsonia*

Terrarium

1. Rain turning to steam
2. Lava
3. Fumaroles

Overview

Beginning with this time-travel session, and using their **Time Travel Journals** to record observations and ideas, students rotate through 10 or more learning stations representing the time period 4.5 BYA–544 MYA. **This period represents almost 90 percent of the entire history of our planet!**

As this period begins, the atmosphere and weather are toxic to life as we know it. The land is lifeless and has no soil, but is experiencing dramatic volcanic activity. Very gradually, over an extremely long period of time, the first plantlike life-forms appear and make the atmosphere oxygen-rich. With oxygen in the atmosphere, many oxygen-breathing organisms evolve; first, single-celled, and later, many-celled. These soft-bodied organisms, lacking any hard body parts, left almost no fossil record for this period—just enough to let us know that such organisms did exist. In this session's activities, early many-celled organisms are represented by the sponge, jellyfish, earthworm, flatworm, and tubifex worm ("blood worm").

In Session 2, student teams circulate and record their observations and ideas at these learning stations:

- Time Travel Aquarium
- Time Travel Terrarium
- Continental Drift
- Fossils
- Organism Adaptations (three kinds)
- "Guts"
- Early Life
- Other Early Invertebrates

At some stations they study individual live organisms that represent life during this period in Earth's history. Over the course of their observations at each station, students develop a sense of how and why life evolved over this period, including the impacts of major events on the planet.

During a wrap-up that pulls the session content together, students and the teacher review each station and discuss the students' ideas about what they learned at each. The teacher adds to their ideas and shares what scientists currently think. The session's information is posted in the

> This session's time-span covers the Precambrian, or Cryptozoic ("Hidden Life") Eon, which includes the Hadean ("Hell-like"), Archaean ("Ancient Life"), and Proterozoic ("Early Life") Eras.
>
> (Although the Hadean was lifeless, it was a time of volcanic and cosmic fury; the "formative years" of our planet!)

beginnings of a **Life through Time** wall chart; the most significant *evolutionary events* are reviewed and placed on a **Class Time Line;** and the period's featured *organisms* are placed on a **Tree of Life** wall chart.

More content is introduced through a dramatic election of the most representative organism of the period, as candidate organisms make "vote-for-me" speeches that recap their importance to the time. Students conclude by making predictions in their journals for what will occur in the next time period.

■ What You Need

Note: Many of the items you gather for this session will also be used in later sessions.

For the teacher:
- ❏ 1 copy of **Fossils—Teacher's Answer Sheets** (pages 64–68; 5 pages total)
- ❏ 1 copy of **"Guts"—Teacher's Answer Sheet** (page 73)

For the class:
- ❏ 1 copy of this session's **Most Representative Organism Script** (pages 69–70)
- ❏ 1 overhead transparency of **"Guts": Single-Celled Organism/Sponge/Jellyfish/Earthworm** (page 72)
- ❏ 1 overhead transparency of **Early Life** (page 74)
- ❏ 1 overhead transparency of **Other Early Invertebrates** (page 75)
- ❏ 1 sheet of butcher paper 7 ft. wide x 6 ft. tall for **Life through Time** wall chart
- ❏ 1 sheet of butcher paper 3 ft. wide x 5 ft. tall for **Tree of Life** wall chart
- ❏ 1 copy of **title sign** for **Life through Time** wall chart (page 76)
- ❏ 1 copy each of these **headers** for **Life through Time** wall chart (page 77):
  - __ Ages
  - __ Aquarium
  - __ Terrarium
  - __ "Guts"
  - __ Continental Drift
- ❏ 1 copy of this session's **Tree of Life Organism Cards** (pages 122–126) to add to the **Tree of Life** wall chart

- ❑ 1 copy of **The Age of** _____ sign (page 78)
  (_Note:_ You may choose instead just to write this information on a blank 8 ½" x 11" sheet of paper—which would allow you to re-use the **Life through Time** wall chart for future classes—or directly on the chart after the election, to simplify.)
- ❑ 1 copy of **Major Evolutionary Events—Time Period #1** (page 82)
- ❑ the adding-machine-tape strip you cut for the next session, "544 MYA–410 MYA," to add to the **Class Time Line**
- ❑ 2 large cloths (such as sheets or large towels) to cover and conceal the aquarium and terrarium until you're ready to show them
- ❑ 1 globe or world map
- ❑ an overhead projector and screen
- ❑ (_optional/recommended_) 1 overhead transparency of **Algae Reproduction** (page 71)
- ❑ (_optional_) laminator, if you wish to protect and reuse the wall chart signs and/or tank illustrations
- ❑ (_optional_) three-hole punch, if you wish the students to keep their **Time Travel Journals** in a three-ring binder

For the Time Travel Aquarium station:
- ❑ 1 aquarium tank (Any size can work, but the larger the tank, the easier it will be to fit increasing numbers of plastic and real animals and plants as the unit progresses. We recommend no smaller than the classic five-and-a-half-gallon size. See the chart of tank dimensions on page 63.)
- ❑ aquarium gravel and/or sand (Coarse gardening sand, sandbox sand, or sand-blasting sand will all work fine. **Note: do not use coral sand or coral gravel;** these are only used for saltwater aquariums and can kill freshwater organisms. Do not use sand or gravel you have collected from an ocean beach, as it will be salty. If your sand or gravel is dusty, rinse it thoroughly before use.)
- ❑ 1 copy of **Aquarium Background—Time Period #1** (page 92A; you may wish to laminate this. _See the chart of tank dimensions on page 63 for helpful information about sizing the background to your tank size.)_
- ❑ enough dechlorinated water to fill and periodically top off the aquarium (You may use bottled spring water, but **do not use distilled water,** because it lacks beneficial minerals. You may also use tap water with dechlorinating liquid, which can be purchased at an aquarium store.)
- ❑ 1 natural (not synthetic) bath sponge (You'll need more for a separate station; see page 41.)
- ❑ 1 (or more) plastic jellyfish (To make your own, see sidebar.)

To make a "jellyfish," cut the bottom off a clear plastic bottle to use as the "dome" of the jellyfish, or use a small clear plastic baggie inverted and stuffed with shimmery plastic. Cut strips of stiff plastic to hang below it for tentacles.

For the Time Travel Terrarium station:

❑ 1 aquarium tank (Again, the larger the tank, the more diorama elements can be added. See the chart of tank dimensions on page 63.)

❑ enough lava dirt (preferable) or lava rock to make a 2- or 3-inch-deep layer in the terrarium (If you cannot find these, you can use other rocks. **Soil is a last resort—there *was* no soil on Earth in the period covered in this session!**)

❑ 1 copy of **Terrarium Background—Time Period #1** (page 92B; you may wish to laminate this. *See page 63 for help in sizing the background to your tank size.*)

Optional: There are several options for making your terrarium "volcanoes" more exciting; you may choose one or all of these effects, depending on your supplies and time.

1. For erupting volcanoes:

❑ 2–3 glass or plastic chemistry vials or tubes, plus one extra to use for adding vinegar during the activity (see "Resources," page 307)

❑ enough baking soda to fill two or three vials two-thirds full

❑ enough chocolate powder to add 1 tablespoon to each volcano eruption mixture, for better appearance

❑ enough red food coloring to add to the volcano eruption mix, for better appearance

❑ 2–3 plastic or foam cups, each with a hole punched in its bottom, to place over the vials to create the slopes of the volcanoes

❑ enough brown clay for sticking pieces of volcanic rock to the rim of the vials or the sides of the cups, for a more realistic appearance

❑ enough vinegar to occasionally add a few drops to each vial

2. For glowing volcanoes:

❑ 1 or more small red bicycle lights with a steady shining option, to make the volcanoes "glow" from within

❑ same number of sealable baggies, to protect the light(s) from rubble and moisture

❑ same number of smaller, cone-shaped clear plastic or foam cups, each with its bottom cut off, to create the slopes of the volcanoes

❑ enough brown clay for sticking pieces of volcanic rock to the sides of the cups, for a more realistic appearance

3. For smoking volcanoes:

- ❏ 3 glass or plastic chemistry vials or tubes (see "Resources")
- ❏ 3 clear plastic or foam cups to invert over vials to create the slopes of the volcanoes; can also just mound the lava rock around the vials
- ❏ enough water to fill vials two-thirds full
- ❏ a few small pieces of dry ice to add to the water in the vials (see "Resources")
- ❏ enough brown clay for sticking pieces of volcanic rock to the sides of the cups, for a more realistic appearance

For the remaining core stations:

- ❏ 2 copies of the station sheet **Early Life** (page 74)
- ❏ 2 copies of the station sheet **Other Early Invertebrates** (page 75)
- ❏ 2 copies of the station sheet **"Guts": Single-Celled Organism/ Sponge/Jellyfish/Earthworm** (page 72)
- ❏ 2 copies of the station sheet **Fossils—Time Period #1** (page 83)
- ❏ 3 copies of the station sheet **Continental Drift—Time Period #1;** two for the station and one for the **Life through Time** wall chart (page 84)
- ❏ several containers to "house" organisms at the stations (Use whatever containers seem appropriate for the organism, such as an old food storage container for flatworms.)
- ❏ 2 or more natural bath sponges
- ❏ algae (Can be obtained from a local birdbath or pond; "scum" or film on the inside of a fish tank or other body of fresh water. May look fuzzy, or like green hair. Seaweed can also be used.)
- ❏ 5 or so earthworms (Can be obtained from the ground or compost pile. Can also be purchased at bait shops or biological supply houses; see "Resources.")
- ❏ 1 water spray bottle, to keep the earthworms moist
- ❏ paper towels, to provide cover for the earthworms

For a class of 32 students, you'll need 10 or more stations for them to rotate through for each of Sessions 2–6. These are the 10 core stations for this session (several will occur in all time-travel sessions):

1) Time Travel Aquarium
2) Time Travel Terrarium
3) Continental Drift
4) Fossils

5) Organism Adaptations: Animal (Sponge)
6) Organism Adaptations: Animal (Earthworm)
7) Organism Adaptations: Plant (Algae)
8) "Guts"
9) Early Life
10) Other Early Invertebrates

In addition to these 10, for this session you may choose to add one or more organism adaptations stations from the following list.

For the optional additional stations:
❑ 2 copies of the station sheet **Algae Reproduction** (page 71)
❑ 2 or more flatworms (Also known as *Planaria*. Can sometimes be collected in freshwater streams and ponds, if you know what you're looking for, or purchased from some aquarium stores or biological supply houses. See "Resources.")
❑ a handful of tubifex worms (Also known as "blood worms." Can be purchased from local aquarium store or biological supply houses; see "Resources.")

Planaria

For each student:
❑ personalized **Time Travel Journal** cover from the previous session (page 27)
❑ personalized journal page labeled **First Life on Earth** from the previous session (page 28)
❑ personalized, stapled **Organism Key** from the previous session (6 pages)
❑ 1 set of **Time Travel Journal** pages labeled **Time Period #1** (pages 85–93; 9 pages total) to add to each student's journal *(Don't copy these until you've added this session's organisms to the **Organism Adaptations** page; see "Getting Ready," below.)*
❑ either a large binder clip or a three-ring binder, for the **Time Travel Journal**

■ **Getting Ready**

Note: **Much of the following preparations will be contributing to the overall set-up of the unit, so while it may seem overwhelming, know that there will be far less for you to prepare in future sessions!**

Before the Day of the Activity

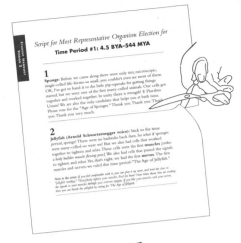

1. Copy and cut up this session's **Most Representative Organism Script** (pages 69–70) so each character's part is on a separate sheet. Depending on your student's comfort with dramatic reading, you may choose one of the following ways to deliver the speeches: have different students read some or all of the parts; read each part yourself; alternate roles with another adult; have other adults or older students visit the class to read the parts.

2. Write the names of the organisms you've acquired for this session in the left-hand column on the **Organism Adaptations** page of the students' journals.

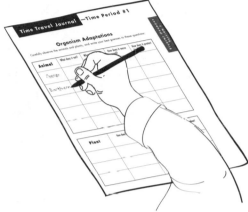

3. Copy the **Time Travel Journal** pages marked **Time Period #1** (pages 85–93; 9 pages total). If using three-ring binders for the journals, punch holes in the pages.

4. Copy and staple the **Fossils—Teacher's Answer Sheets** (pages 64–68). These are the answer sheets for all time-travel sessions. Copy the **"Guts"—Teacher's Answer Sheet** (page 73).

5. Copy and set aside the overhead transparencies for this session: **Algae Reproduction; "Guts": Single-Celled Organism/ Sponge/Jellyfish/Earthworm; Early Life;** and **Other Early Invertebrates** (pages 71–72 and 74–75).

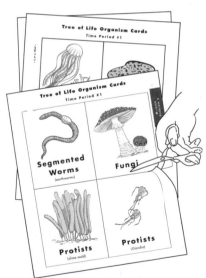

6. Label the containers you'll be using for the organisms at the **Organism Adaptations** stations. Invertebrate containers should be labeled "Invertebrate" and also more specifically, "earthworm" and "sponge." Plant container should be labeled "Plant" and also "algae."

7. Copy and cut up the **Tree of Life Organism Cards** (pages 79–81) for this session, ready to place on the class **Tree of Life** wall chart.

8. Set up the **Time Travel Aquarium.**

   a. Cover the bottom of the aquarium with a thin layer of aquarium gravel or sand.

   b. Slowly add spring water or dechlorinated tap water to the aquarium until it's approximately two-thirds full.

   c. Tape **Aquarium Background—Time Period #1** (page 92A) to the back of the aquarium. Don't use too much tape, as you'll

Erupting Volcano

vinegar drops from vial

baking soda,
chocolate powder and
red food coloring

clay

lava rocks

lava dirt

Glowing Volcano

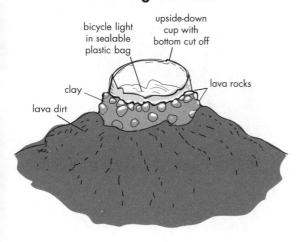

bicycle light
in sealable
plastic bag

upside-down
cup with
bottom cut off

clay

lava rocks

lava dirt

need to remove the background at the end of this session and add
it to the **Life through Time** wall chart.

d. Place one natural sponge in the aquarium. It will probably need a
weight attached to it, such as a binder clip, to prevent it from
floating.

e. Place the plastic jellyfish in the aquarium. By carefully placing it on the
surface, you may be able to get it to float and look more life-like.

9. Set up the **Time Travel Terrarium.**

a. Tape **Terrarium Background—Time Period #1** (page 92B) to
the back of the terrarium, using a minimum of tape. (It, too, will
later be added to the **Life through Time** wall chart.)

b. Cover the bottom of the terrarium with a few inches of lava dirt
or lava rock. (You'll need to either remove this layer and replace it
with soil, or cover it with a layer of soil, when preparing for the
next session.)

c. Prepare the "volcanoes," if you're using them. (See also "On the
Day of the Activity," on page 46.) You may, of course, combine the
effects if you wish.

Erupting volcano option:
1) For each erupting volcano, fill a vial two-thirds full with
baking soda and add 1 tablespoon of chocolate powder and a
little red food coloring to each for an authentic appearance.

2) Bury each vial in a mound of lava dirt with the opening
exposed at the surface; or invert a plastic or foam cup (with a
hole punched in the bottom) over the vial to create the slope
of the volcano.

3) Make a small "lip" of dirt around the opening of each vial to
look like a volcanic crater. Stick brown clay to the exposed
side of each cup and attach bits of lava rock.

4) Set aside a container of vinegar and one empty vial with
which to drop vinegar into the volcanoes.

Glowing volcano option:
1) For each glowing volcano, prepare a clear plastic or foam cup
with the bottom cut off. (This will form the walls of the
volcano from which the red glow emanates.)

2) Cover the sides of the cup with brown clay and attach bits of lava rock to camouflage the cup.

3) Set aside the bicycle light(s), which you'll prepare just before the activity.

Smoking volcano option:

1) For each smoking volcano, fill a vial two-thirds full with water.

2) Bury each vial in a mound of lava dirt with the opening exposed at the surface; or invert a plastic or foam cup (with a hole punched in the bottom) over the vial to create the slope of the volcano.

3) Make a small "lip" of dirt around the opening of each vial to look like a volcanic crater. Stick brown clay to the exposed side of each cup and attach bits of lava rock.

4) Set aside small pieces of dry ice, which you'll drop into the vials to create "smoke" during the activity.

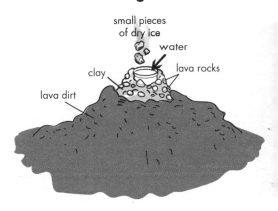

Smoking Volcano

10. Set up the **Life through Time** wall chart and overhead projector.

   a. Choose a location on your class wall or bulletin board for the **Life through Time** wall chart; you'll need an area 7 feet wide x 6 feet tall. Because you'll also be showing overhead transparencies from this spot during the debriefing, the space should accommodate the projector and provide a wall or screen for showing the transparencies.

   b. Attach your butcher paper to the wall and set up the overhead projector.

   c. Draw vertical lines down the paper to create five columns. **Note that the Aquarium and Terrarium columns need to be big enough to hold the aquarium and terrarium background illustrations that are added every day.**

   d. Copy and attach the **title sign** (page 76) to the top of the chart.

   e. Copy, then attach, the following **headers** (page 77) to the tops of the columns: Ages, Aquarium, Terrarium, "Guts," and Continental Drift.

Note: The **Life through Time** wall chart will be filled in **from the bottom up,** with the lowest row representing the most ancient time span. See page 60 to see how the chart will look after this session, and page 217 for its appearance when complete, after all the time-travel sessions.

f. Copy **The Age of** _____ sign (page 78) or write the information on a blank 8 ½" x 11" sheet of paper.

11. Set up the **Tree of Life** wall chart.

a. Choose a location on your class wall or bulletin board for the **Tree of Life** wall chart; you'll need an area approximately 3 feet wide x 5 feet tall.

b. Attach your butcher paper to the wall.

c. Write the title "Tree of Life" across the top of the chart.

12. Copy and set aside the **Major Evolutionary Events—Time Period #1** sheet (page 82); you'll post this over the **Class Time Line** later in the session.

On the Day of the Activity

1. Finish setting up the **Time Travel Terrarium** volcano options from page 44, if you're using one or more of them.

Erupting volcano option:
During class, occasionally add drops of vinegar from the extra vial into the volcanoes to make them "erupt."

Glowing volcano option:
a. For each volcano, turn the red bicycle light to the steady-shining option and place it in a sealable baggie, removing excess air before sealing.

b. Bury the bag in the lava dirt so the light shines upward. Invert the prepared plastic or foam cup over the light to create the walls of the volcano from which the red glow emanates. The light fixture itself should not be visible.

Smoking volcano option:
During class, occasionally drop small pieces of dry ice into the water in each vial to create a volcanic smoking effect. (Be sure to use gloves or tongs when handling dry ice, and don't permit students to touch it themselves.)

2. Set up the **Organism Adaptations** stations.

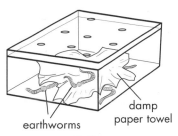

damp
paper towel

earthworms

a. Place the **earthworms** in one or two labeled containers with a damp paper towel, so the worms don't dry out and die. Periodically check and remoisten or lightly spray with water if necessary. Provide another paper towel or other object under which the worms can escape from the light. Set the containers at a station.

b. Place the **sponges** in one or more labeled containers and put them at a station.

sponges

c. Place the **algae** in one or more labeled containers and put them at a station.

3. Make copies of the station sheets and set up the remaining core stations.

pond algae

a. Set out the two copies of **Early Life** (page 74) at a station.

b. Set out the two copies of **Other Early Invertebrates** (page 75) at a station.

c. Set out the two copies of **"Guts": Single-Celled Organism/ Sponge/Jellyfish/Earthworm** (page 72) at a station.

d. Set out the two copies of **Fossils—Time Period #1** (page 83) at a station.

e. Set out two copies of **Continental Drift—Time Period #1** (page 84) at a station.

4. Set up the optional additional station(s).

a. Make two copies of the **Algae Reproduction** station sheet (page 71) and **cut off the bottom (labeled) halves.** Set out the two *top halves* at a station.

and/or

b. Place your **flatworms** in one or two labeled containers with spring or dechlorinated water and put them at a station.

and/or

c. Place some of the **tubifex worms** in two labeled containers with spring or dechlorinated water and put them at a station. (A turkey baster works well for picking up and transporting tubifex worms.) Add the others to the **Time Travel Aquarium.**

5. Gather the students' **Time Travel Journal** covers, journal page labeled **First Life on Earth,** and **Organism Keys** from Session 1, ready to distribute.

6. Put out sets of **journal pages** labeled **Time Period #1.** Have the binder clips or three-ring binders ready to distribute to students.

7. Set aside the third copy of the **Continental Drift—Time Period #1** sheet to add to the **Life through Time** wall chart.

8. Cover the aquarium and terrarium with the cloths so the students can't see them until you're ready for the unveiling.

9. Have the adding-machine-tape strip "544 MYA–410 MYA" ready to add to the **Class Time Line.**

 ■ **Introducing the Time Travel Journal**

1. Distribute the binder clips or three-ring binders to each student, depending on how you'd like them to put together their **Time Travel Journals.**

2. Distribute the **Time Travel Journal** covers, **Organism Keys,** and **First Life on Earth** pages from the previous session.

3. To each student distribute a set of **Time Travel Journal** pages labeled **Time Period #1.**

4. Have students clip their journals together or put them in three-ring binders. (Don't let them use staples; pages will be added to the journals each session.)

5. Let the students know they'll begin their first investigation of a time period long ago. Show them the strip you added to the Class Time Line at the end of the last session. Tell them that the time period extends from the beginning of the Earth, 4.5 billion years ago, to 544 million years ago—an astonishing 3 billion, 956 million year stretch. Let them know that this period represents nearly 90 percent of the entire history of our planet.

4.5 BYA—544 MYA

---

6. Tell them that activities have been set up at stations around the room to help them understand what life and the environment were like at that time. Their job will be to visit each station, record their observations, and answer questions in their journals. Point out that pages in their journals correspond to the different stations.

■ Preparing for Time Travel

Go over each of the station's activities by pointing out the station materials and providing the following stations overview. **Ask students to be sure to leave the station illustrations where they are** when it's time to rotate to a new station. Note that the basic structure of the stations will be the same for the next four sessions, so at this point it's worth making sure students understand and are performing the tasks at each station. In the next four sessions, station explanations can be much more brief.

■ Stations Overview

Stations 1 and 2: Time Travel Aquarium and Terrarium stations

1. Tell students that when they arrive at each of these stations, they should observe the dioramas and draw pictures of them in their journals.

2. Tell them to use their **Organism Key** to try to identify life-forms in the aquarium and terrarium, and to circle any they can identify (including organisms represented by plastic figures and illustrations) on the Organism Key.

3. Tell them they will label the organisms on their diorama drawings.

4. If you're using dry ice in the volcano, make sure students know they are **not** to touch it.

Station 3: Continental Drift

1. Tell students they'll draw in their journals the continents as they appeared during this time period, referring to the illustration provided at the station.

2. Point out the weather and atmosphere report, which will help students understand how life was changing at this time. (It really engages the students if you read a sample report like the one on the next page, delivering it as if you were a weather reporter.)

**Stations recap—
Session 2/Time Period #1**

1) Time Travel Aquarium
2) Time Travel Terrarium
3) Continental Drift
4) Fossils
5) Organism Adaptations:
   Animal (Sponge)
6) Organism Adaptations:
   Animal (Earthworm)
7) Organism Adaptations: Plant
   (Algae)
8) "Guts"
9) Early Life
10) Other Early Invertebrates

Optional:
• Algae Reproduction and/or
• Flatworm and/or
• Tubifex worm

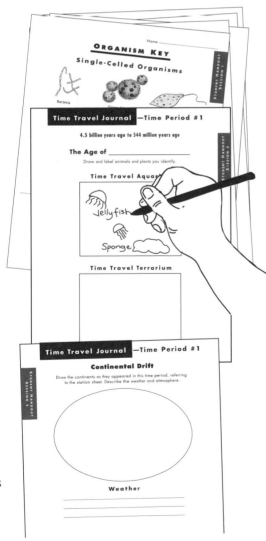

Weather
Expect it to be hot, with chance of volcanoes.
Volcanic dust, thunder, lightning, and constant rain.
Swimming not advisable—even the seas are hot!

Atmosphere
Hydrogen, carbon monoxide, ammonia, and methane.
Time Traveler advisory: You would die in these poisonous gases!

Station 4: Fossils

Tell students they'll examine pictures of fossils at this station and try to figure out what kind of animals or plants they once were. Basing their guesses on life-forms in their **Organism Key,** students can write in their journals any fossils they manage to identify. (**Say that some are difficult—so if students can identify even a few, they're doing great!**)

Stations 5 and 6 (plus two of the optional stations): Organism Adaptations: Animals

1. Refer to the **Organism Adaptations** page of the students' journal and ask students to look at the Animals chart.

2. Show students a living organism at one or more of the stations. Tell them they'll be working and thinking like wildlife biologists, carefully observing each organism and making guesses about the following:

   - what it eats
   - what eats *it*
   - how it moves around
   - how it protects itself

   Be sure to model and emphasize that *all* organisms must be treated gently and with respect.

3. Let students know not to worry if they don't know the correct answer for any of the questions, but say they need to think about them and record their ideas. Some answers may be obvious after watching the organisms; others may take more thinking and guessing. Tell them to write down their best guesses.

Station 7: Organism Adaptations: Plant

1. Refer to the **Organism Adaptations** page of the students' journal and ask students to look at the Plants chart.

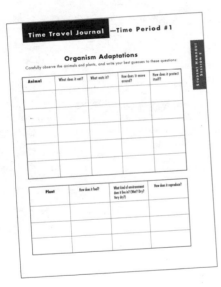

We've included a set of teacher's answer sheets (pages 64–68) to identify the fossils from each of the five time-travel sessions. This is merely a resource in case students and/or you are curious about a particular fossil.

2. Tell students they'll touch the plant at the station and describe how it feels. Then they'll try to guess whether it lives in a wet, dry, or very dry environment and how it might reproduce. (To learn why algae are really "plantlike," not true plants, see "Background for the Teacher" on page 270.)

Station 8: "Guts"

1. Refer to the **"Guts"** page of the students' journals.

2. Ask students which organism on the sheet is a **single-celled** life-form. [The amoeba.]

3. Tell students they will draw the path they think food takes as it travels through the digestive systems of the amoeba, the sponge, the jellyfish, and the earthworm.

4. Tell them to then number the four organisms from **simplest** digestive system (as #1) to **most complex** (#4), using their best guesses.

Stations 9 and 10: Early Life and Other Early Invertebrates

1. Tell students they'll be looking at the pictures of *early life* at station 9 and should note what the organisms have in common. Ask students to write two or more statements in their journals about general similarities and differences between these organisms. [Possible answers include "they're microscopic"; "many are single-celled"; "they're rounded"; "they appear transparent."]

2. Tell students that at station 10 they should look at the pictures of *other early invertebrates,* draw the same kinds of conclusions about similarities and differences among these organisms, and write two or more statements about them in their journals. ["They're multi-celled"; "they often occur in colonies," etc.]

Recommended optional additional station: Algae Reproduction

1. Ask students what they know about how plants reproduce (most students know about seeds, at least). Accept all answers.

2. Tell the students that at this station, their challenge will be to think about how this plant might reproduce, and then write down their best guesses.

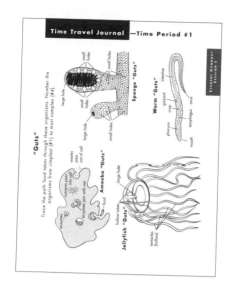

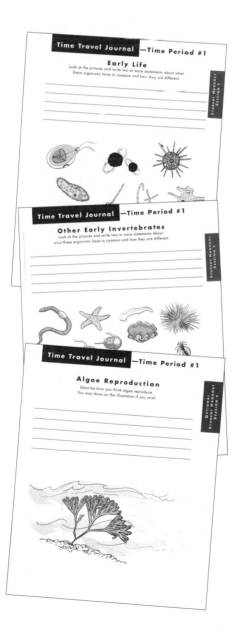

■ Overview Wrap-Up

1. When you've finished the overview of all the stations, let students know that at the end of each day they'll be asked to make predictions about the next session's time period, based on what they learned.

2. To add a little intrigue, tell the class that one of the living organisms they'll study over the course of the unit is a "killer"! Its kind were responsible for killing off most of the first organisms on Earth. Reassure them that it will pose no danger to them—and that, in fact, without it humans and most other life-forms on Earth never would have evolved into what they are today. Have them try to figure out which organism you're referring to, but don't reveal the answer! [The killer was early single-celled plants, which produced oxygen and transformed the atmosphere, killing off all the anaerobic life-forms on Earth. Students will hear about it in one of the campaign speeches, later in the session.]

■ Traveling through the Stations

1. Divide students into teams of two.

Students may not need as much time for this station rotation as in future sessions, because there were so few types of organisms during this period on Earth.

2. Assign student teams to their first stations (up to two teams per station), then let them rotate through the others at their own pace. **There's no specified order to visiting the stations.** Advise them that when they rotate, they should go to stations that have few people at them and avoid stations that are crowded. (They can return to a crowded station later when fewer students are there.) If any teams finish all the stations earlier than others, they can return to stations of their choice for further observations. Remind students to leave the station sheets where they found them, for other pairs to use.

Remember, means you can pause here if you're teaching multiple classes.

3. During rotations, be sure to attend to the Time Travel Terrarium station and "fire off" the volcanoes a few times—adding vinegar and/or dry ice as necessary. If you're using the red bicycle light inside a volcano, you may want to dim the classroom lights initially, to add to the visual effect.

■ Time Travel Debrief

Note: A large amount of information has been included in each station's wrap-up discussion, below. This information is NOT meant to simply be read to the students. The overall goal is to inspire students to really **think** about these ideas—so, as much as possible, information should be brought out through open-ended questions directed at the students, and through teacher-guided discussion.

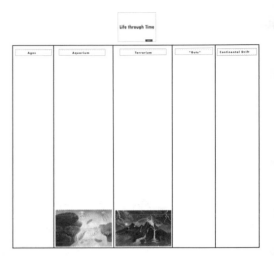

1. Seat the students so they can see the **Life through Time** wall chart for a review of the stations. They'll need their **Time Travel Journals** and **Organism Keys** for the discussion.

2. Remove the background illustrations from the Time Travel Aquarium and Terrarium and tape them on the **Life through Time** wall chart under the "Aquarium" and "Terrarium" headings. Refer to the illustrations to point out organisms during the discussion.

3. Debrief the **Aquarium** and **Terrarium** stations.

   Ask a few time travelers to share what they noticed about the Earth during this time period, including:

   - land
   - water
   - plants
   - animals
   - weather
   - atmosphere
   - continents

Note: If you were obliged to use soil as a substrate in this session's Time Travel Terrarium, explain to students that there actually **was no true soil** during this period. Soil is created by eroded rock mixing with the decay and excretions of terrestrial animals and plants, and no animals or plants yet existed on land.

4. Debrief the **Early Life** and **Other Early Invertebrates** stations.

   a. Place the overhead transparency **Early Life** on the overhead projector. Ask your students to turn to the Early Life page in their journals, and ask a few students to share their statements about early life during this period.

b. During the discussion, be sure the following points are brought up about early life on Earth:

Early Life on Earth

- **It was small.** Life-forms were microscopic at first, starting as single-celled, then evolving into many-celled organisms.

- **Many lived in colonies.** Before multi-celled organisms evolved, many single-celled life-forms lived close together to get a lot of what they needed in life. (The waste products of their neighbors, for instance, could provide nutrients they couldn't make for themselves.)

- **It was simple.** An important point throughout the unit is the evolution from simplicity to complexity. The single-celled organisms were very simple. The jellyfish is many-celled and much more complex than earlier life-forms, but still very simple when compared with other organisms, such as mammals, which evolved later.

- **The first organisms were soft.** The earliest animals didn't have hard shells, exoskeletons (skeletons on the outside), or endoskeletons (skeletons on the inside).

- **Life was aquatic.** There was no life on land.

- **Life-forms were anaerobic.** The very first life didn't use oxygen, because there was none. Over time, early single-celled plants began to photosynthesize, and over millions of years made an oxygen-rich atmosphere. The oxygen was deadly to the earliest organisms, but new, oxygen-breathing organisms flourished.

---

c. Show the **Other Early Invertebrates** transparency and ask students to continue their observations about life during this period.

d. Ask a few students to refer to their journals and share their statements about what the organisms from the **Early Life** station had in common and how they were different.

e. Have students do the same with the organisms from the **Other Early Invertebrates** station.

Technically speaking, earthworms did not appear until after the first true soil formed on Earth. We use it here for consistency.

5. Debrief the **Organism Adaptations** stations.

a. Revisit the animal and plant adaptation stations, asking the students for a few of their ideas on each organism's adaptations. If possible, do this by placing each organism in front of the class. For the algae station, ask some of your students how they answered the questions about whether algae live in a wet, dry, or very dry environment.

b. Feel free to add anything more you know about the organisms. Again, the small facts are not as important as the general trends, and the primary goal of these stations is to get students to think about an organism's **adaptations** to its environment.

6. Debrief the **"Guts"** station.

a. Show the overhead transparency **"Guts": Single-Celled Organism/Sponge/Jellyfish/Earthworm.**

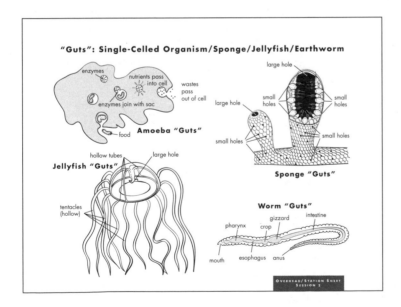

While we've chosen the amoeba to illustrate a "two-way traffic" feeding system (food entering and leaving from the same or similar locations), it's important to note that the amoeba is NOT part of a lineage that eventually led to the one-way feeding systems of, say, the earthworms. We've chosen this familiar protozoan simply to stand in contrast to the other systems. What matters is that students recognize different feeding strategies and modifications in strategies through time.

b. Referring to your teacher's answer sheet if needed, explain that single-celled organisms like the amoeba take in food through any part of the cell membrane in a little "pocket" in the cell that becomes a food sac inside the organism. The nutrients are digested

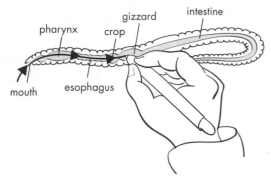

Worm "Guts"

and absorbed into the cell from the food sac, and then wastes are released the same way the food came in: through the cell membrane. Ask the students to imagine what it would be like if people digested food this way!

There are many different types of worms. The worm represented in the "Guts" illustration is a segmented worm (earthworm), which has the one-way-traffic gut. The flatworm, however, doesn't possess a gut. It takes food in and eliminates wastes through the same opening.

c. Ask a few students to share their ideas about food pathways in the other three organisms (sponge, jellyfish, and earthworm), or have them come up and draw arrows on the transparency to show the way food is digested in each organism.

d. Explain to students that before the earthworm came along, food entered and waste exited from the same opening(s). The earthworm's "one-way traffic" system (one way in and a different way out) was a huge breakthrough.

Here's one way to compare the two kinds of systems:

• The two-way system in the sponge and jellyfish creates something like a traffic jam inside the organism. Food and wastes get mixed together. All the cells in the linings have to be able to "do it all": absorb, digest, and excrete.

• The earthworm's two-opening, one-way system is more like one-way traffic. The digestive tract (gut) can specialize, with separate parts responsible for food intake, digestion, absorption, and elimination of wastes. The gut is similar to a tunnel going though an organism's body; the food source enters at one end and exits at the other. The food and wastes are kept separate. The body takes the nutrients it needs from the food moving through this tunnel.

e. You may choose to point out specific structures in the earthworm's gut:

• mouth—where it takes in food
• pharynx—in early organisms, an extension of the mouth
• esophagus—where the food is delivered to the stomach
• crop—storage area
• gizzard—muscular organ where tiny particles of sand help grind up food
• intestine—where enzymes help complete digestion and nutrients are absorbed into the worm's body
• anus—where undigested wastes are eliminated

f. Add the **"Guts": Single-Celled Organism/Sponge/Jellyfish/ Earthworm** station sheet (page 72) to the bottom of the **Life through Time** wall chart, under the "Guts" heading.

Worm "Guts"

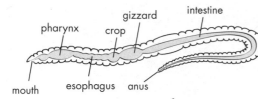

7. Debrief the **optional Algae Reproduction** station.

a. Ask your students how they think algae reproduce.

b. Show the overhead transparency **Algae Reproduction.** Explain that algae simply expel their reproductive cells out into the water. The water keeps them moist and mixes them around so male and female reproductive cells can connect.

c. Ask the students how this strategy might relate to the kind of environment (dry, wet, or very wet) algae lives in. [The algae depends on being surrounded by water.]

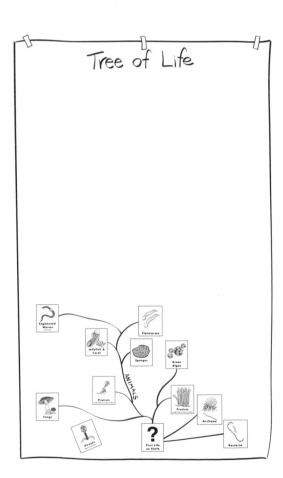

■ Tree of Life Wall Chart

At this point in the stations debrief, pause to fill in the **Tree of Life** wall chart. (In subsequent sessions, adding to the **Tree of Life** chart can be delegated to students and/or done between classes.)

1. Have students recall the "trees of life" they made in the previous session with their **Student Tree of Life Cards.** Explain that those are similar to the **Tree of Life** chart the class will be assembling each session. It represents how scientists think organisms are related, using the most up-to-date information.

2. Using this session's **Tree of Life Organism Cards,** place the pictures on the **Tree of Life** wall chart as shown, discussing with the students as you go.

This is how the Tree of Life wall chart will look at the end of this session.

To see how the completed chart will look at the end of the whole unit, see page 184.

The virus illustration includes question marks because scientists don't agree on whether viruses are alive or not—let alone if and how they're related to (other) organisms! This is because they're not made of cells, but of biochemical genetic material (DNA or RNA) surrounded by a protein "shell." They can only multiply when a cell they've invaded (a host cell) reproduces—and takes them along for the ride.

■ Time Travel Debrief Continued

1. Debrief the **Continental Drift** station.

    a. Point out the continents as they appear today on the globe or world map, and then the picture of the continents as they appeared during this session's time period.

    b. Ask the students to report what they noticed about the continents and atmosphere at the station. They may have noticed that:

- As the period began, there were no "continents," just colliding islands.
- What will eventually become Africa, South America, and North America are at the equator.
- It's extremely cold, and glaciers cover much of the world.

    c. Add the station sheet **Continental Drift—Time Period #1** to the **Life through Time** wall chart under the heading "Continental Drift."

2. Debrief the **Fossils** station.

    a. Ask if students were able to guess any of the fossils at the station. Be sure to ask what their reasoning was for identifying the organism(s) they did.

    b. Referring to your **Fossils—Teacher's Answer Sheet** for this time period, compare students' guesses with the real identifications. Be sure to emphasize that the correctness of their guesswork is far less important than the reasons they gave for figuring it out.

    c. If no students guessed at any fossils (and depending on time and interest), you may identify some or all of the fossils at the station from your answer sheet.

■ Major Evolutionary Events

1. Ask students to consider all they've just discussed, then brainstorm some major evolutionary changes that took place in the organisms or their habitats during the period they just studied. Refer to the **Major Evolutionary Events—Time Period #1** sheet as a content checklist. Introduce any significant events not brought up by your students.

Major Evolutionary Events—Time Period #1

- **First life on Earth—single-celled organisms arrive on the scene.** These first life-forms can survive in early Earth's oxygen-free atmosphere and weather. (In fact, oxygen would be toxic to them!)

- **First plantlike organisms evolve.** Over millions of years, through photosynthesis, they convert Earth's atmosphere to become oxygen-rich. The smallest algae are extremely important organisms; they will become one of the basic foods on which all marine life depends.

- **The first many-celled organisms evolve.** The sponge is one of the first.

- **By the end of the period, soft-shelled marine invertebrates are common.**

- **Flatworms develop the first "eyes" (photoreceptors) and the beginnings of a brain.**

- **Segmented worms develop the first one-way digestive system.**

- **Jellyfish-like animals develop the first nerves and muscles.**

2. Ask students how their observations compare with the predictions they made about this period in Session 1.

3. Post the sheet **Major Evolutionary Events—Time Period #1** above this session's strip of the **Class Time Line**.

■ Most Representative Organism of the Age

Now that the class has a sense of what life was like during this time period, it's time to "name the age" based on what organism the students think was most representative of the time.

1. Ask students what organisms might be considered for Most Representative Organism of this time period. Accept all answers. Let them know that several candidates will be presenting their cases, and that the class will get to vote after each candidate has made a brief "campaign speech." Students are also welcome to improvise a speech for a worthy candidate they feel may have been left out.

2. Conduct the campaign.

   a. Hand out the cut-up speeches to students or adults who will play each role.

   b. Have the volunteers read their speeches, in numerical order. Be sure to introduce and write the name of each candidate on the board before the speech begins.

1

Sponge: Before we came along there were only tiny, microscopic, single-celled life-forms-so small, you couldn't even see most of them. OK, I've got to hand it to the little pip-squeaks for getting things started, but we were one of the first many-celled animals. Our cells got together and worked together. In unity there is strength! E Pluribus Unum... the only candidate that helps you at bath time. Please... Thank you. Thank you. Thank you...

2

Jellyfish (Arnold Schwartzenegger voice): Stick to the time period, sponge! There were no bathtubs back then. So *what* if sponges were many-celled-so were we! But we also had cells that worked together to tighten and relax. These cells were the first **muscles** *[strikes a body builder muscle flexing pose]*. We also had cells that passed the signals to tighten and relax. Yes, that's right, we had the first **nerves**. The first muscles and nerves-we ruled this time period-"The Age of Jellyfish."

Note to the actor: If you feel comfortable with it, you can play it up more, and lead the class in "jellyfish aerobics." "Everybody tighten your muscles. Feel the burn! Now relax. You are sending the signals to your muscles through your nervous system. If you like your muscles and your nerves, then you can thank the jellyfish by voting for 'The Age of Jellyfish.'"

c. When all speeches have been given, you may want to ask if students have any other organisms from this period to nominate and make an improvised campaign speech for.

3. Hold the election.

a. After the speeches, reconvene the class to vote for the most representative organism.

b. Explain that each student may only vote once. Tell them to vote for what they truly believe is the **most representative organism** of the period they just studied—not just for their friends who read the speeches! Review the possibilities with the class:

- The Age of Sponges
- The Age of Jellyfish
- The Age of Worms
- The Age of Algae
- The Age of Single-Celled Life

Tell students they may also vote for a name that would include them all:

- The Age of Early Life

c. Conduct and tally the vote. Write the name of the winner on the **Age of_____** sign, a blank 8 ½" x 11" sheet of paper, or directly on the **Life through Time** wall chart. Have students write it on the aquarium and terrarium drawing page of their **Time Travel Journal.** If you made a sign, add it to the wall chart, under the "Ages" heading.

Note: The chart should now be arranged as shown here.

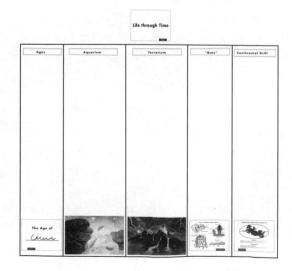

■ Class Time Line

1. If there's time, take a moment with the class to appreciate the major events of the period, referring to the sheet above the **Class Time Line.** Ask for a few students' ideas about how these events affect life on Earth to this day. [For one thing, we—and every living thing we live with and depend on—exist!]

2. Attach to the **Class Time Line** the strip of adding-machine tape you pre-cut for the next session, "544 MYA–410 MYA." Tell your students the next session will focus on this period in Earth's history.

3. Write the years of the next time period on the board, using any or all of the following forms. The first two are very impressive looking; the last is the simplest way to write it (and a form scientists use). It can be fun and instructive for students to see all three:

   - 544 million–410 million years ago
   - 544,000,000–410,000,000 years ago
   - 544–410 MYA

4. Have your students write in their **Time Travel Journals** their predictions about the organisms and habitat changes they think they'll encounter in the next period, including land, water, plants, animals, weather, atmosphere, and continents.

5. Ask students to bring in organisms you'll need for the next session, to be used at the organism adaptation stations.

If possible, find a location in your classroom to display interesting organisms, such as flatworms, for students to continue observing during the duration of the unit.

■ Going Further

1. Incorporating Related Activities
Any activities you have from other curricula relating to microorganisms, cells, worms, algae, photosynthesis, early invertebrates, volcanoes, beginning of life on Earth, oxygen, digestive system, nervous system, eyes, or muscles would be excellent additions at this point.

2. Algae Play
See the script for *The Algal Invasion,* on pages 94–98. Make enough copies for four or more actors, and have your students play the parts in front of the class.

3. Songs
The flamenco song *Gusano (I Am A Worm)*—lyrics on page 321—makes an engaging and energetic "Going Further" to this activity.

(You can preview the song at www.moo-boing.com. The site also provides information on how to purchase this and other songs mentioned in this guide.)

A general note on songs:

The songs at the back of this guide are by the educational rock band The Bungee Jumpin' Cows. They are included as part of a multiple-intelligence approach to the subject matter, and for reinforcement of concepts. Many teachers of upper elementary and middle school are not accustomed to incorporating songs into science lessons, largely because there are very few educational songs geared for this age group. Many teachers have found that The Bungee Jumpin' Cows' songs have the "edginess" and musicality their students enjoy.

For a fuller classroom experience, see www.moo-boing.com for information on how to purchase these songs on cassette or CD. There are many ways to incorporate songs into lessons, but probably the easiest is simply to have students read along to lyrics on individual sheets or an overhead transparency as they listen to the song. Keeping songs available for individual listening (with a portable CD player, for instance) or playing them during station activities allows for the repetition often needed with music. Many teachers also have individual or small groups of students perform a song for the class.

Sizing Your Time Travel Aquarium and Terrarium Background Illustrations

A FEW STANDARD TANK CAPACITIES AND DIMENSIONS

TANK CAPACITY	TANK DIMENSION (H x L)
5 $\frac{1}{2}$ gal.	10" x 16"
10 gal.	12" x 20"
15 gal.	12" x 24"
20 gal.	16" x 24"

HOW TO SIZE THE ILLUSTRATIONS TO MATCH YOUR TANK SIZE

For this size tank...	and this size original illustration...	reduce (▼) or enlarge (▲) by...	to get this size background illustration for your tank.
5 $\frac{1}{2}$ gal. (10" x 16")	11" x 17"	▼ 91%	10" x 15 $\frac{3}{8}$"
10 gal. (12" x 20")	11" x 17"	0%	11" x 17"
15 gal. (12" x 24")	11" x 17"	▲ 109%	12" x 18 $\frac{1}{2}$"
20 gal. (16" x 24")	11" x 17"	▲ 142%	15 $\frac{5}{8}$" x 24"

Fossils—
Teacher's Answer Sheet

for Session 2

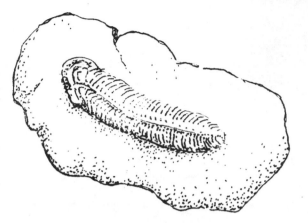

Spriggina

Early Soft Coral

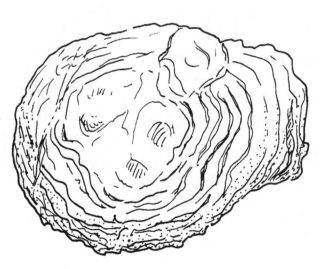

Early Jellyfish

Arthropod Tracks

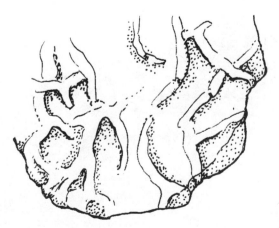

Worm Burrows

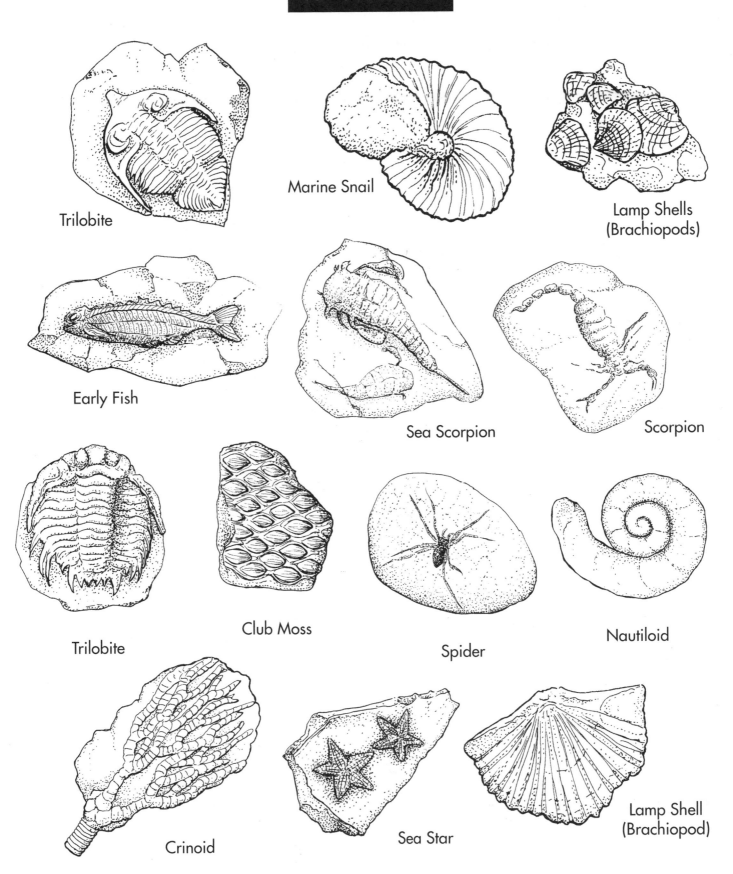

Trilobite

Marine Snail

Lamp Shells
(Brachiopods)

Early Fish

Sea Scorpion

Scorpion

Trilobite

Club Moss

Spider

Nautiloid

Crinoid

Sea Star

Lamp Shell
(Brachiopod)

Fossils—
Teacher's Answer Sheet
for Session 4

Coral

Early Sea Urchin

Fern

Early Bony Fish

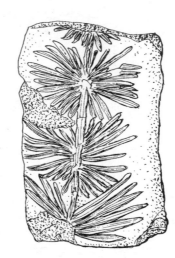

Horsetail

Sequoia

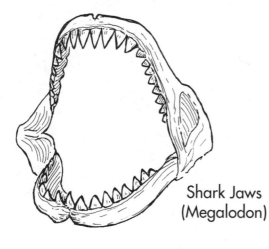

Shark Jaws
(Megalodon)

Ginkgo

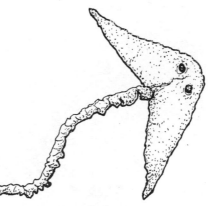

Armored Fish
(head only)

Amphibian
(Diplocaulus)

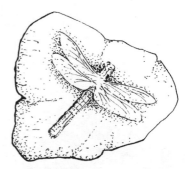

Dragonfly

Fossils—
Teacher's Answer Sheet

for Session 5

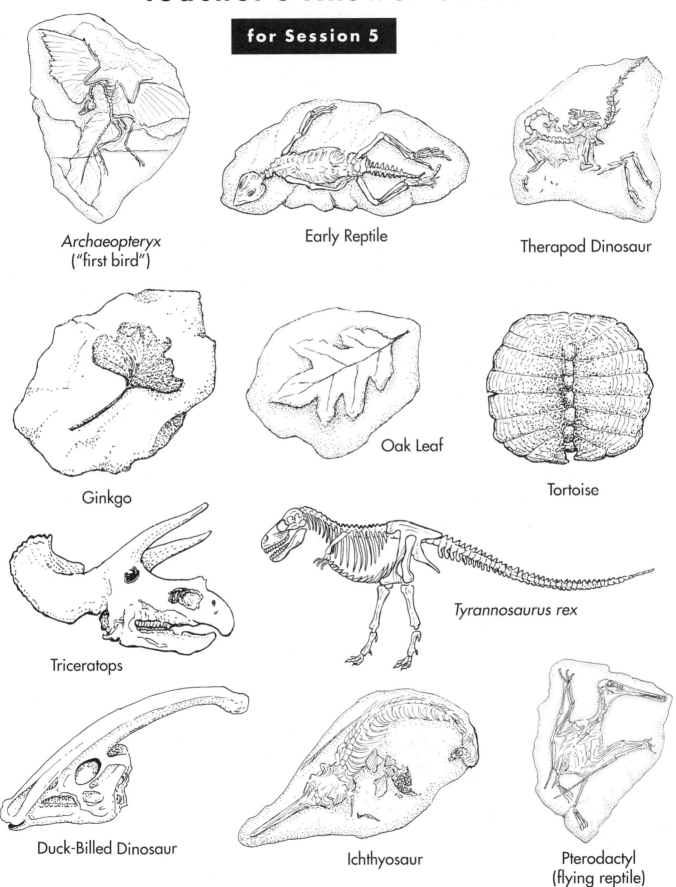

Archaeopteryx
("first bird")

Early Reptile

Therapod Dinosaur

Ginkgo

Oak Leaf

Tortoise

Triceratops

Tyrannosaurus rex

Duck-Billed Dinosaur

Ichthyosaur

Pterodactyl
(flying reptile)

Fossils—
Teacher's Answer Sheet

for Session 6

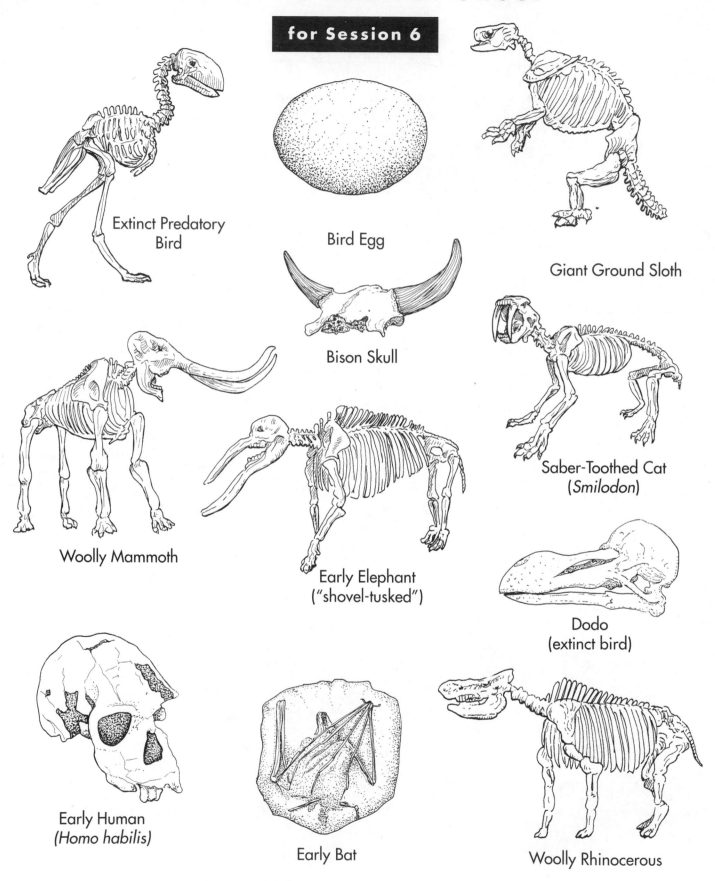

Extinct Predatory
Bird

Bird Egg

Giant Ground Sloth

Bison Skull

Woolly Mammoth

Early Elephant
("shovel-tusked")

Saber-Toothed Cat
(*Smilodon*)

Dodo
(extinct bird)

Early Human
(*Homo habilis*)

Early Bat

Woolly Rhinocerous

Script for Most Representative Organism Election for
Time Period #1: 4.5 BYA–544 MYA

1

Sponge: Before we came along there were only tiny, microscopic, single-celled life-forms—so small, you couldn't even see most of them. OK, I've got to hand it to the little pip-squeaks for getting things started, but we were one of the first **many-celled animals.** Our cells got together and worked together. In unity there is strength! *E Pluribus Unum!* We are also the only candidate that helps you at bath time. Please vote for the "Age of Sponges." Thank you. Thank you. Thank you. Thank you very much.

2

Jellyfish (Arnold Schwartzenegger voice): Stick to the time period, sponge! There were no bathtubs back then. So *what* if sponges were many-celled—so were we! But we also had cells that worked together to tighten and relax. These cells were the first **muscles** *[strikes a body builder muscle flexing pose].* We also had cells that passed the signals to tighten and relax. Yes, that's right, we had the first **nerves.** The first muscles and nerves—we ruled this time period—"The Age of Jelly-fish."

Note to the actor: If you feel comfortable with it, you can play it up more, and lead the class in "jellyfish aerobics." "Everybody tighten your muscles. Feel the burn! Now relax them. You are sending the signals to your muscles through your nervous system. If you like your muscles and your nerves, then you can thank the jellyfish by voting for 'The Age of Jellyfish.'"

3

Worm: Nerves and muscles are nice, but there's something else about the jellyfish you wouldn't like. It's disgusting. They ate and pooped through the same opening. You can thank us worms for evolving **two openings,** one for eating, the other for pooping. And what about flatworms? They didn't have these two openings, but they did evolve some of the first beginnings of eyes (which could detect light), and beginnings of a brain. We gave the world **a new digestive system, brains,** and **eyes!** It should be called "The Age of Worms."

4

Algae: Yeah, but none of you could have lived without us. Before we came along, the atmosphere and weather were poisonous for all of you *[points at other candidates].* Through photosynthesis, we made **oxygen,** which you *[points at class]* breathe. We changed the Earth completely. We are the most important organism *ever* on Earth. Even in modern times, we still make most of the oxygen on Earth. We are also the only candidate that makes its own food from gases and the sun. This should be called the "Age of Algae"!

5

Single-Celled Animal: There is a killer in this room! And the killer is right *[points at algae]* there! Killers! Killers! Before you came along there were many single-celled animals that *didn't* need oxygen. In fact, the oxygen you made poisoned them, and killed most of them off! That's when single-celled oxygen breathers, and later all *you* oxygen breathers *[points at animal candidates and class]* evolved. Single-celled animals were the first life on Earth! We deserve to have this period named the "Age of Single-Celled Life."

Algae Reproduction

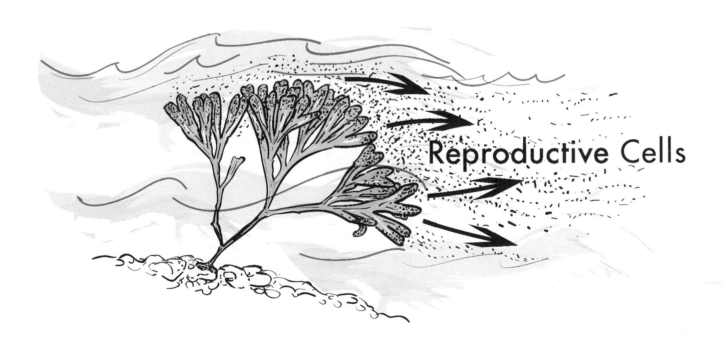

Reproductive Cells

"Guts": Single-Celled Organism/Sponge/Jellyfish/Earthworm

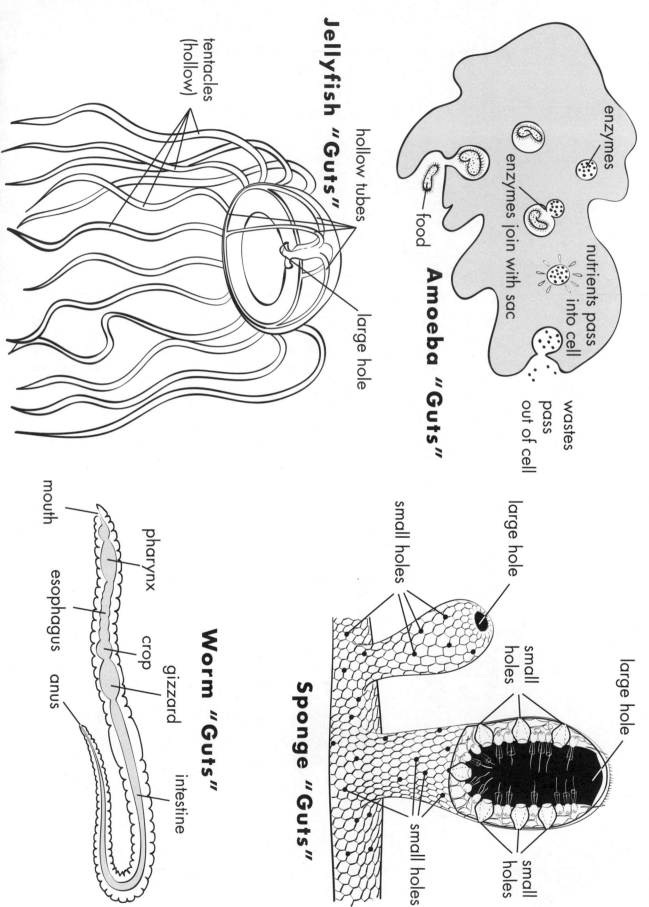

Amoeba "Guts"

enzymes

enzymes join with sac

nutrients pass into cell

food

wastes pass out of cell

Jellyfish "Guts"

tentacles (hollow)

hollow tubes

large hole

Sponge "Guts"

small holes

large hole

large hole

small holes

small holes

small holes

Worm "Guts"

mouth

pharynx

esophagus

crop

gizzard

anus

intestine

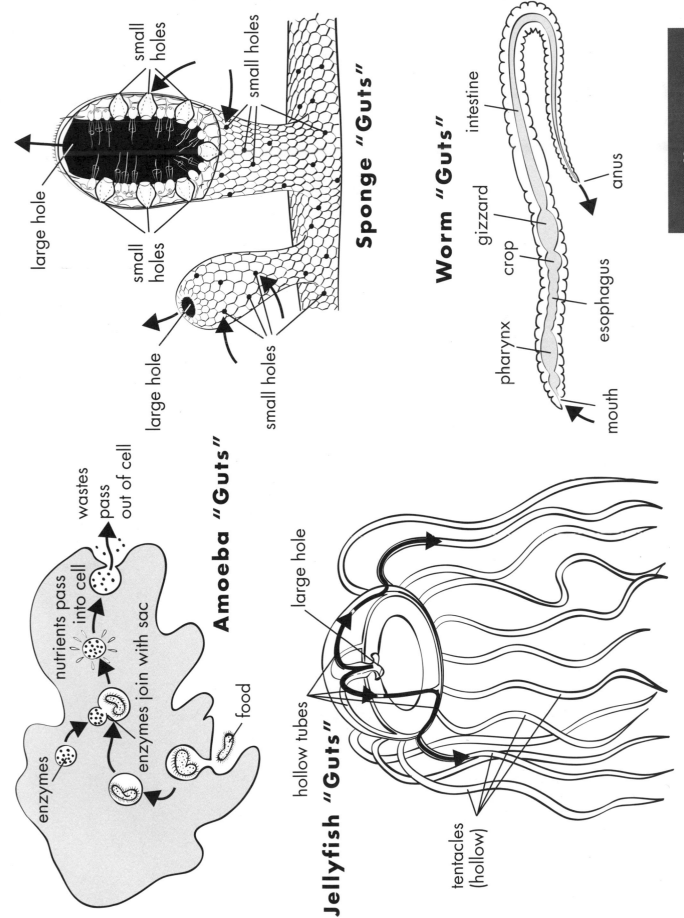

"Guts" — Teacher's Answer Sheet

Sponge "Guts"

small holes

small holes

large hole

small holes

large hole

small holes

Amoeba "Guts"

wastes pass out of cell

nutrients pass into cell

enzymes join with sac

enzymes

food

Jellyfish "Guts"

large hole

hollow tubes

tentacles (hollow)

Worm "Guts"

intestine

anus

gizzard

crop

esophagus

pharynx

mouth

Early Life

Note: All are microscopic

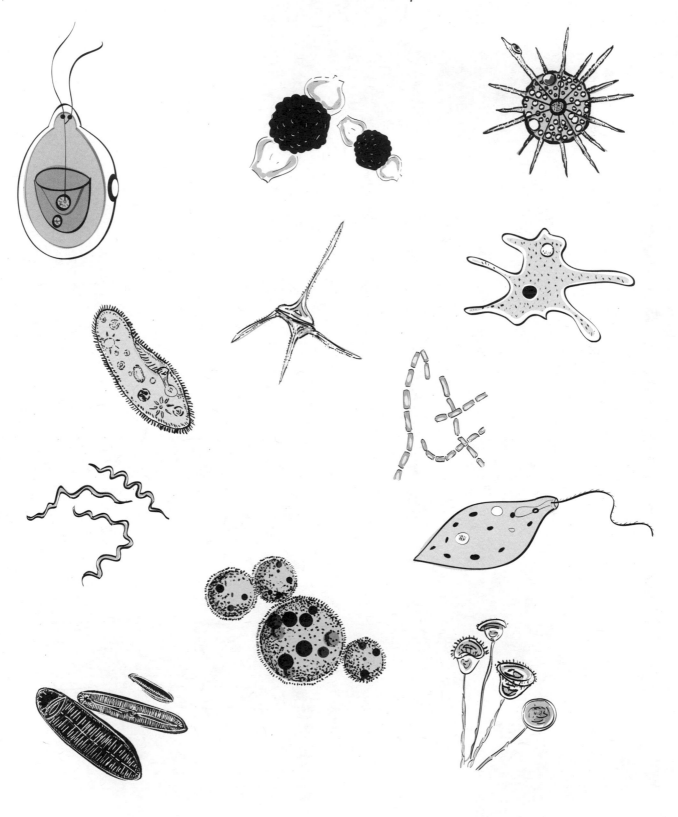

Other Early Invertebrates

Life through Time

Ages

Aquarium

Terrarium

"Guts"

Continental Drift

The Age of

Tree of Life Organism Cards

Time Period #1

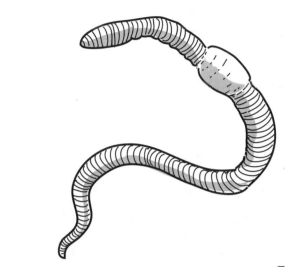

Segmented Worms
(earthworms)

Fungi

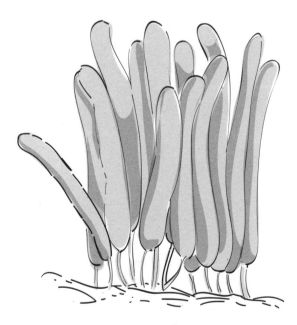

Protists
(slime mold)

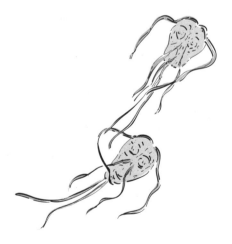

Protists
(*Giardia*)

Jellyfish & Coral

Sponges

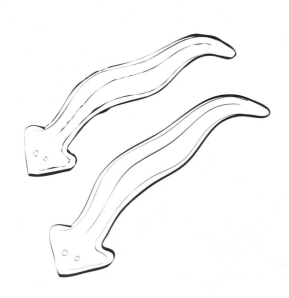

Flatworms
(Planaria)

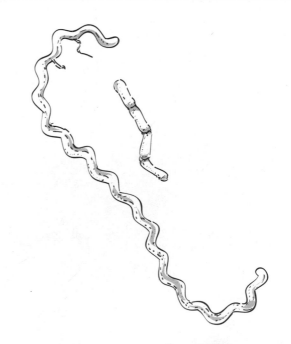

Bacteria

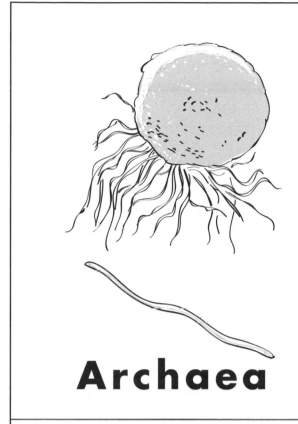

Archaea

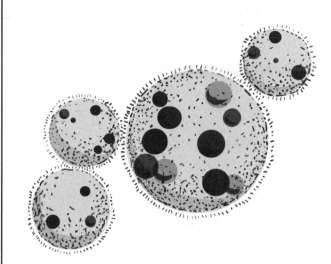

Green Algae

First Life on Earth

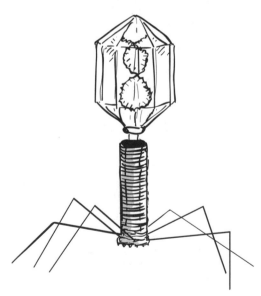

Viruses
(plant? animal? other?)

Major Evolutionary Events—Time Period #1

- **First life on Earth—single-celled organisms arrive on the scene.** These first life-forms can survive in early Earth's oxygen-free atmosphere and weather. (In fact, oxygen would be toxic to them!)

- **First plantlike organisms evolve.** Over millions of years, through photosynthesis, they convert Earth's atmosphere to become oxygen-rich. The smallest algae are extremely important organisms; they will become one of the basic foods on which all marine life depends.

- **The first many-celled organisms evolve.** The sponge is one of the first.

- **By the end of the period, soft-shelled marine invertebrates are common.**

- **Flatworms develop the first "eyes" (photoreceptors) and the beginnings of a brain.**

- **Segmented worms develop the first one-way digestive system.**

- **Jellyfish-like animals develop the first nerves and muscles.**

Fossils—Time Period #1

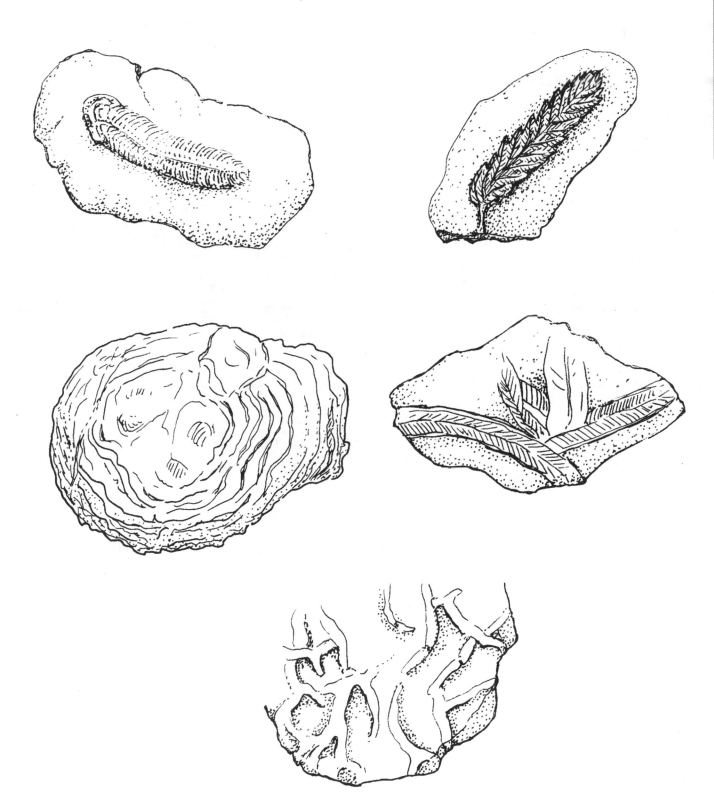

Continental Drift—Time Period #1

4.5 BYA–544 MYA

EQUATOR

Weather

At first: Expect it to be hot, with chance of volcanoes. Volcanic dust, thunder, lightning, and constant rain. Swimming not advisable—even the seas are hot!

Later: We have a definite cooling trend. Worldwide glaciers will make temperatures a bit nippy.

Atmosphere

At first: Hydrogen, carbon monoxide, ammonia, and methane. **Time Traveler advisory:** You would die in these poisonous gases!

Later: Mostly nitrogen and oxygen. Time Traveler advisory has been lifted. You could survive these gases.

4.5 billion years ago to 544 million years ago

The Age of _____

Draw and label animals and plants you identify.

Time Travel Aquarium

Time Travel Terrarium

© 2003 The Regents of the University of California. May be duplicated for classroom or workshop use.

Fossils

Label any fossils you're able to identify.
(Some are challenging!)

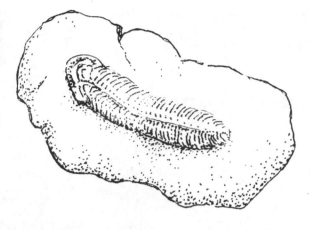

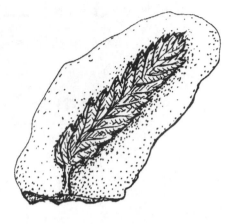

_____

_____

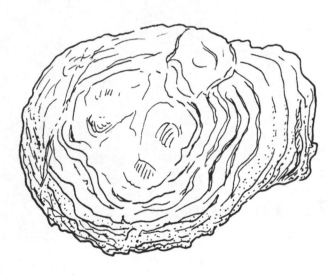

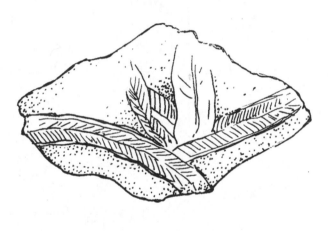

_____

_____

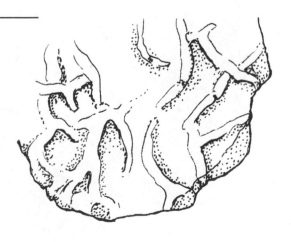

_____

Early Life

Look at the pictures and write two or more statements about what
these organisms have in common and how they are different.

_____

_____

_____

_____

_____

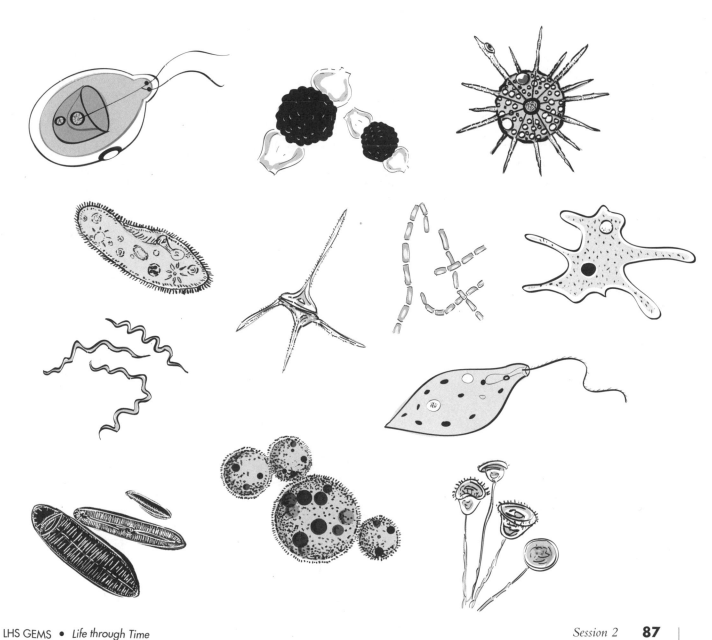

Other Early Invertebrates

Look at the pictures and write two or more statements about
what these organisms have in common and how they are different.

_____

_____

_____

_____

_____

Organism Adaptations

Carefully observe the animals and plants, and write your best guesses to these questions:

Animal	What does it eat?	What eats it?	How does it move around?	How does it protect itself?

Plant	How does it feel?	What kind of environment does it live in? (Wet? Dry? Very dry?)	How does it reproduce?

"Guts"

Trace the path food takes through these organisms. Number the organisms from simplest (#1) to most complex (#4).

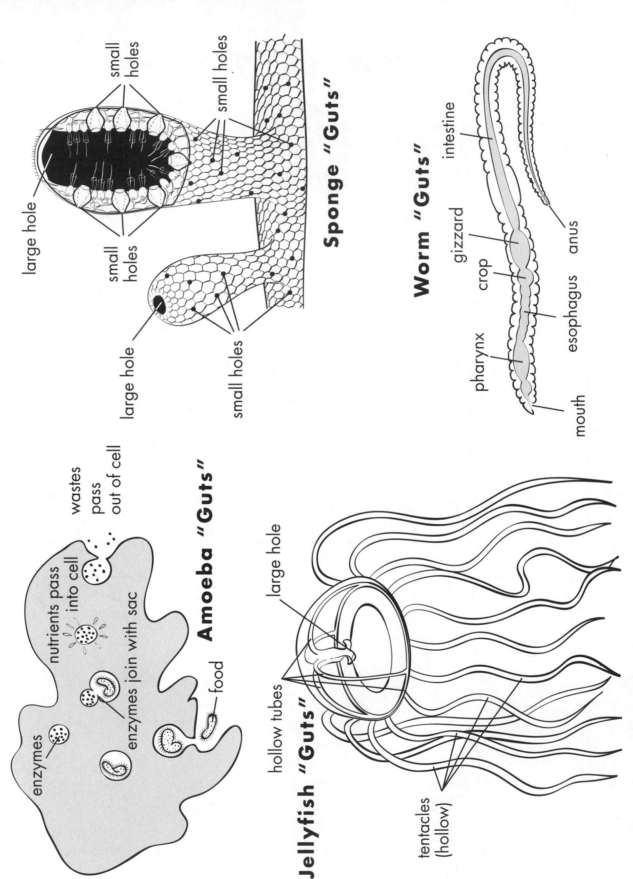

Sponge "Guts"

Worm "Guts"

intestine

gizzard

crop

pharynx

esophagus   anus

mouth

small holes

small holes

large hole

small holes

large hole

small holes

large hole

Amoeba "Guts"

wastes pass out of cell

nutrients pass into cell

enzymes

enzymes join with sac

food

Jellyfish "Guts"

large hole

hollow tubes

tentacles (hollow)

Continental Drift

Draw the continents as they appeared in this time period, referring to the station sheet. Describe the weather and atmosphere.

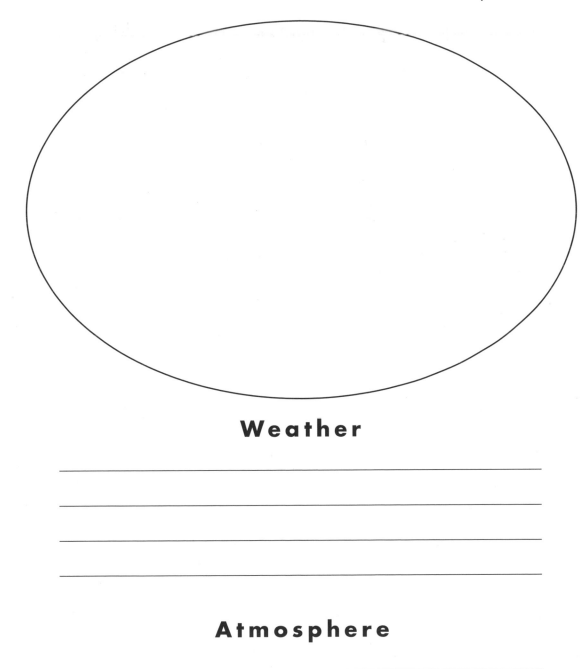

Weather

_____

_____

_____

_____

Atmosphere

_____

_____

_____

_____

Predictions for Next Time Period

Write your predictions for how organisms, land, water, weather, and atmosphere
may change in the next time period.

_____

_____

_____

_____

_____

_____

_____

_____

_____

_____

_____

_____

_____

_____

_____

_____

LHB '02

Algae Reproduction

Describe how you think algae reproduce.
You may draw on the illustration if you wish.

_____

_____

_____

_____

_____

OPTIONAL STUDENT HANDOUT SESSION 2

The Algal Invasion

Pronunciation for the teacher:
   Alga is the singular term, and is pronounced "al-gah."
   Algae is the plural term, and is pronounced "al-gee."

The Scene: Freshwater during The Age of _____

Characters:

General alga—Like a military general. Authoritative, brusque when speaking to mutant algae, and in charge.

Mutant alga—A little whiney. A smart mouth. Always bringing up the different perspective.

Other alga—A lower-ranking alga.

Narrator

Narrator: Algae gather preparing for an assault on land. General alga is addressing the troops.

General alga: All right men gather round here.

Mutant alga: We're not men.

General alga: OK gather around algae. Now listen up.

Mutant alga: We can't listen—we're algae. And you can't talk and we can't think or plan so what's the point of this?

General alga: Look it's a skit about evolution, OK. And no, algae can't talk, think or plan, and yes evolution is all about chance—not planning, but if we had an algae skit and acted like real algae, it would be boring. And if we did things the way they really happened, the skit would be billions of years long, so give me a break. Now who are we?

Everyone: We're algae!

Mutant alga: How come it says "alga" in the script then? Shouldn't we all be yelling, "We're alga!"?

General alga: Algae is plural, you mutant. Alga is singular. So who are we?

Everyone: We're algae!

Mutant alga: *Yells* I'm alga!

General alga gives Mutant alga a stern look.

Mutant alga: But *we're* algae! Right on algae! Go algae! Algae rule!

General alga: That's right, we're algae, and we're gonna take over the world. We're already everywhere in the waters, but why stop there?

Mutant alga: But we're *just* algae, we're nothing.

General alga: We're nothing?? Nothing? Before we came along the atmosphere was full of carbon dioxide. Then we came along. What did we do? We made oxygen. We took water, carbon dioxide, and energy from the sun, and we photosynthesized. We made our own food—sugars and starches. We made oxygen. Over millions of years we made so much oxygen that 20% of Earth's atmosphere is oxygen.

Mutant alga: But we're murderers! The oxygen we created killed off most of the bacteria that lived here before us. We practically massacred life on Earth!

General alga: Yes, you're right again mutant. Most of the bacteria that lived on Earth before us were killed off by the oxygen. But we opened things up for oxygen-breathing bacteria, and their descendents that have now taken over. We have helped give them life. We did this. Algae have changed the world. We're everywhere in the seas, which cover most of this planet. But we can't stop here. The land is ours for the taking. There is nothing living on the land.

Mutant alga: Uh excuse me, but isn't there like a reason that there's nothing living on the land? Like isn't it *dry* on land? I'm wet, I like being wet, and if I'm *not* wet, then I'm gonna die.

General alga: Yes it's dry, but it's pretty moist on the shores. And there's nothing else there. Nothing that could eat us.

Mutant alga: Nothing that could eat us *yet*. Uh but isn't there like not nutrients too? I like life here in the water. I get to float around and photosynthesize all day. I get minerals from the waters around me and I make food from the sun, water and gases. I got no complaints.

General alga: There are minerals on land too mutant! The whole land is made up of minerals. Where do you think the minerals we have in the water come from? And you want sun to make food? There's even more sun when you're not under water.

Mutant alga: We can't survive out there though. We're not adapted for it. We'll die!

Everyone: Oh no! We're all going to die!

General alga: Yes most of us will die. But out of millions of us there's got to be some that will be better at living out of the water than the others, and some of those may survive long enough to reproduce. And over millions of years our children's children's children's children's children's children's children's children's...

Mutant alga: *Interrupts* You've made your point.

General alga: Anyway, they will someday conquer even the driest parts of the land.

Mutant alga: Yeah, but we'll be dead. What do we care?

General alga: We're algae, and we don't care. But it's gonna happen sooner or later. As algae we can't be proud either, but as long as we're in this skit, we can be proud. We're algae! Nobody's ever going to thank us for this but we'll do it anyway. In fact people will always look down on us. They'll call us names like kelp, pond scum, water plant, allergy...

Mutant alga: It's not allergy, you said it's algae.

General alga: Yeah, I know, but some people will get the words mixed up.

Mutant alga: People?

General alga: An advanced animal that won't evolve for billions of years. They will depend on us completely, but they'll never appreciate us. They'll breathe the oxygen we make. They won't be able to make food from sunshine and gases like we can, so the only food they'll have will be us plants, and animals that eat plants. Or animals that eat animals that eat animals that eat plants. Or...

Mutant alga: What's an animal?

General alga: The descendents of those inferior bacteria that can't make their own food.

Mutant alga: Dang free-loaders! Parasites!

General alga: Yes, but there's one thing that perhaps those incompetent non-photosynthesizing people-animals of the future should be most grateful to us for.

Mutant alga: What's that?

General alga: Amazing stuff.

Everyone: What?

General alga: Ice cream. Stuff from algae will make ice cream smooth. Other stuff too, like toothpaste, soup, salad dressing, whipped cream, beer foam, syrup, jam, jelly, pudding, sauces, gravy. But the greatest of all these will be ice cream.

Mutant alga: Ice cream? What's it like? Sunshine?

General alga: No, it's not important for survival, like the sun, but they'll eat it for fun.

Mutant alga: What's fun?

General alga: We're algae, and we'll never have it, but people will.

Mutant alga: It's not fair.

General alga: No it's not. They'll never thank us for it either. They'll call us other names like carageenan, agar and alginate, because they'll be embarrassed to write algae on the ingredients list of their packages. But the worst name of all they'll use most often. They will call us seaweed.

Mutant alga: Say it isn't so!

General alga: It will be so.

Mutant alga: But how are we gonna do this? We can't float on land, so what are we gonna do?

General alga: We'll have to evolve stiff stalks to hold us upright.

Mutant alga: We're gonna dry out, so what are we gonna do?

General alga: We'll have to evolve a watertight skin.

Mutant alga: Our spores will dry out, so what are we gonna do?

General alga: We'll have to evolve a system that won't dry out—seeds. But we've got to stop talking and take action. None of us will have any of these things anyway, but our descendants will. Let's get ready. Give me an "A."

Everyone: "A"

General alga: Give me an "L."

Everyone: "L"

General alga: Give me a "G."

Everyone: "G"

General alga: Give me another "A."

Everyone: "A"

General alga: Give me an "E."

Everyone: "E"

General alga: Now what's that spell?

Everyone: Algae!

General alga: What's that spell?

Everyone: Algae!

Mutant alga: What's that *smell?*

Everyone: Algae!

General alga: Now let's hit the beach!

Narrator: Later on, General alga is dying. Another alga reports on the results of the assault.

General alga: What's the report on our assault?

Other alga: Billions of casualties sir. The beaches are full of the smell of rotting dead algae.

General alga: Well we knew the losses would be high. But hasn't anyone survived?

Other alga: Only a few. Some mutants that had a little thicker skin have been able to survive long enough to reproduce.

General alga: *Surprised* Mutants?

Other alga: Yes, some of them. Of course most of the mutations were not helpful, and they died off quicker than anyone. But there are some, and their spores have survived in moisture along the shore, so they will reproduce, but we'll have to do better.

Everyone: *Begin humming "America the Beautiful" in the background.*

General alga: Someday we shall. We'll all soon be gone, but someday there will be plants all over the land. There will be plants hundreds of feet high, some that will live thousands of years. And they will feed all kinds of animals by converting carbon dioxide gas and water to food with the energy from the sun. And in the seas I foresee that we will continue to be the most important life on Earth. We will make most of the oxygen, and we will be the basis of the ocean's food web. And they'll owe it all to algae!

The end (or the beginning)…

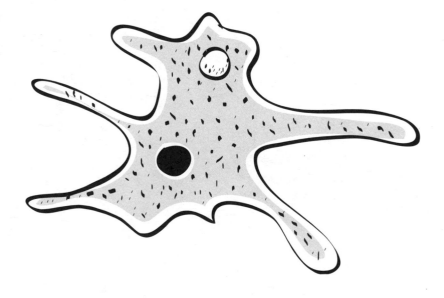

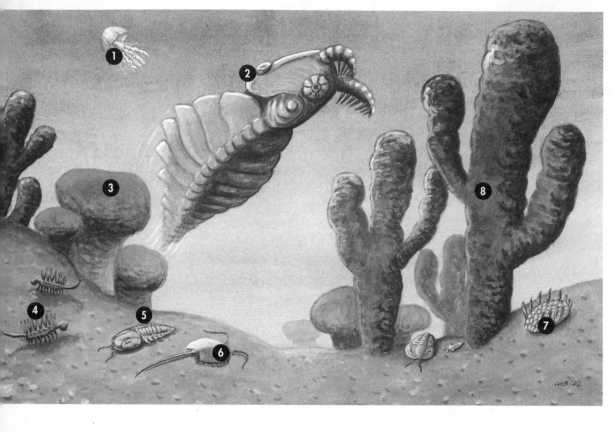

Aquarium

1. Jelly
2. *Anomalocaris*
3. Stromatolites
4. *Hallucigenia*
5. Trilobites
6. *Burgessia*
7. *Wiwaxia*
8. Sponges

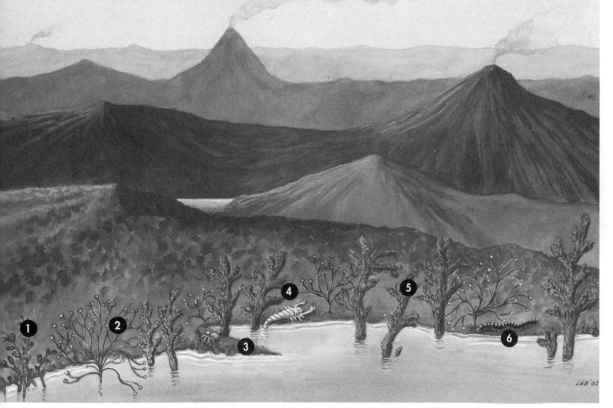

Terrarium

1. *Zosterophyllium*
2. *Cooksonia*
3. Spider
4. Early Scorpion
5. *Asteroxylon*
6. Millipede

Overview

In this session students follow a similar series of activities to those in Session 2, but focus on a new time period: 544–410 MYA.

As this age begins, the first plantlike organisms and animals have colonized the land. Their waste products and decayed bodies are creating the first true soil. Most significantly and dramatically, the seas are experiencing an explosive appearance of hard-shelled arthropods and shellfish, which will leave a great quantity of evidence in the fossil record. (So phenomenal was this evolutionary development that biologists and geologists call this period the "Cambrian Explosion.") Trilobites (ancient marine arthropods) are at their peak, and the first animals with the foundations of backbones (sea squirts and early fish) are also beginning to appear in the seas. By the end of this period, plants and animals will have evolved to survive on land in substantial numbers.

As before, the students rotate through a series of stations, including the Time Travel Aquarium and Terrarium, Fossils, Organism Adaptations, and Continental Drift. There are new stations on Early Land Plants and Arthropods. Students use their **Time Travel Journals** and **Organism Keys** to record observations and ideas about adaptation, and to attempt to identify the organisms they see.

During the wrap-up discussion, the focus is on invertebrates and land plants. Again, the session's accumulated information is added to the **Life through Time** wall chart, the period's most significant evolutionary events are reviewed and placed on the **Class Time Line,** and organisms featured during the session are added to the **Tree of Life** wall chart. The class elects the most representative organism of the age, and once again the students make predictions for what may occur in the next time period.

> This session represents the Cambrian, Ordovician, and Silurian Periods of the Paleozoic ("Ancient Life") Era of the Phanerozoic ("Evident Life") Eon.

> *Take a moment to fully appreciate the incredible diversity of arthropods that made the great move to colonize land in this period. Almost 80 percent of all named animal species are arthropods—mostly insects. Arthropods are to this day ubiquitous; they live around us (crabs and bees), on us (mites), and occasionally thanks to us (mosquitoes)!*

■ What You Need

For the teacher:

❑ the appropriate **Fossils—Teacher's Answer Sheet** for this session (page 65), from your set

For the class:

❑ 1 copy of this session's **Most Representative Organism Script** (pages 117–118)

- ❑ 1 overhead transparency of **Moss Reproduction** (page 119)
- ❑ 1 overhead transparency of **Early Land Plants** (page 120)
- ❑ 1 overhead transparency of **Arthropods** (page 121)
- ❑ 1 overhead transparency of **From Worm to Insect** from Session 1 (page 25)
- ❑ 1 overhead transparency of **Algae Reproduction** from Session 2 (page 71)
- ❑ 1 overhead transparency of **Other Early Invertebrates** from Session 2 (page 75)
- ❑ 1 copy of this session's **Tree of Life Organism Cards** (pages 122–126) to add to the **Tree of Life** wall chart
- ❑ 1 copy of **The Age of** _____ sign (page 78) (Remember, you may instead write this information on a blank 8 ½" x 11" sheet of paper or directly on the **Life through Time** wall chart after the election, to simplify.)
- ❑ 1 copy of **Major Evolutionary Events—Time Period #2** (page 127)
- ❑ the adding-machine-tape strip you cut for the next session, "410 MYA–286 MYA," to add to the **Class Time Line**
- ❑ the **Time Travel Aquarium** and **Time Travel Terrarium** from the previous session
- ❑ the two large cloths from the previous session, to cover the aquarium and terrarium
- ❑ an overhead projector and screen

For the Time Travel Aquarium station:
- ❑ 1 copy of **Aquarium Background—Time Period #2** (page 136A)
- ❑ enough dechlorinated water to top off the aquarium, if needed
- ❑ freshwater snail(s) (Can be purchased in aquarium stores or found in freshwater ponds and streams.)
- ❑ aquatic insects (Can be caught with a net in a local pond or stream.)
- ❑ sea urchin shell(s)
- ❑ plastic sea star(s)
- ❑ plastic mollusk(s), or real shells such as clam or mussel shells

For the Time Travel Terrarium station:
- ❑ 1 copy of **Terrarium Background—Time Period #2** (page 136B)
- ❑ enough soil to make a 2- or 3-inch-deep layer in the terrarium
- ❑ at least one of the following early plants:
  - __ moss (Can be obtained from yards, or purchased in bags from a nursery or hardware store.)

___ liverwort (Can be obtained from a nursery or garden; it's an herb with broad heart-shaped leaves.)

___ club moss (Ground pine or princess pine are examples of club mosses living today.)

❑ plastic scorpion(s)

❑ plastic millipede(s)

❑ live land snail(s) (*Note:* If the land snails are kept in an open container, they may escape and eat nearby paper products...such as unlaminated background illustrations—yikes! They'll also eat your plants.)

For the remaining core stations:

❑ 2 copies of the station sheet **Early Land Plants** (page 120)

❑ 2 copies of the station sheet **Arthropods** (page 121)

❑ 2 copies of the station sheet **Fossils—Time Period #2** (page 128)

❑ 3 copies of the station sheet **Continental Drift—Time Period #2;** two for the station and one for the **Life through Time** wall chart (page 129)

These are the six core stations for this session:
1) Time Travel Aquarium
2) Time Travel Terrarium
3) Continental Drift
4) Fossils
5) Early Land Plants
6) Arthropods

In addition to these six, you'll need at least four organism adaptations stations from the following list. (The stations above involve mostly paperwork, so the addition of living-organism stations will make the session much more interesting.)

For the additional four or more stations:

❑ several containers to "house" organisms at the stations (Use whatever containers seem appropriate for the organism.)

❑ *Triops* (tadpole shrimp) eggs, hatched into adults (It takes a few days to a week before they're large enough to observe. See "Making It as Easy as Possible" on page 8, and "Resources" on page 307.)

❑ isopods—pill bugs (roly-polies) and sow bugs

❑ hermit crab★

Scientists think Triops evolved 245 million years ago, so to be totally accurate it doesn't belong in this session's time period. If you're uncomfortable with this inaccuracy, you may include them during Session 5 instead of Session 3. We've chosen to include them here because they represent the kinds of organisms that did live during the period but are now extinct. They're an excellent example of an early aquatic arthropod.

- ❏ crayfish ("crawdad")
- ❏ freshwater snails
- ❏ centipede
- ❏ millipede
- ❏ spider
- ❏ clam
- ❏ sea star★
- ❏ silverfish
- ❏ moss
- ❏ liverwort
- ❏ club moss (ground pine or princess pine are living examples of early land plants)
- ❏ invertebrate exoskeleton (shed!)
- ❏ trilobite fossil

★These organisms in particular are delicate creatures and require careful handling and special care. In many states (including California) it is ILLEGAL to remove organisms from the wild.

Note: If you have more than one individual of a type of organism (two spiders or two isopods, for example), each individual can be its own station. Ideally, however, one type of organism should not be used for more than two stations.

For each student:
- ❏ **Time Travel Journal** from the previous session
- ❏ **Organism Key** from the previous session
- ❏ 1 set of **Time Travel Journal** pages labeled **Time Period #2** (pages 130–136; 7 pages total) to add to each student's journal *(Don't copy these until you've added this session's organisms to the **Organism Adaptations** page; see "Getting Ready," below.)*

■ Getting Ready

One Week Ahead

If you've decided to use them, acquire the *Triops* eggs (see "Resources" on page 307), so they'll have time to hatch out before the day of the activity.

Before the Day of the Activity

1. Copy and cut up this session's **Most Representative Organism Script** (pages 117–118) so each character's part is on a separate sheet.

If other adults are reading the parts, contact them to schedule their time.

2. Write the names of the organisms you've acquired for this session in the left-hand column on the **Organism Adaptations** page of the students' journals.

3. Copy the **Time Travel Journal** pages marked **Time Period #2** (pages 130–136; 7 pages total), punching holes in them if students are using three-ring binders.

4. Copy and set aside the overhead transparencies for this session: **Moss Reproduction, Early Land Plants,** and **Arthropods** (pages 119–121).

5. Label the containers you'll be using for the organisms at the **Organism Adaptations** stations. They should first be labeled "Arthropods" or "Plants." Under these general names, list more specific names too, such as "spider" or "moss."

6. Copy and cut up this session's **Tree of Life Organism Cards** (pages 122–126), ready to place on the **Tree of Life** wall chart.

7. Update the **Time Travel Aquarium.**

   a. Top off the aquarium with spring or dechlorinated water, if needed.

   b. Tape **Aquarium Background—Time Period #2** (page 136A) to the back of the aquarium, using a minimum of tape. (Remember, it'll come down at the end of the session to be added to the **Life through Time** wall chart.)

   c. Add the sea urchin shell(s), plastic sea star(s), plastic mollusk(s), and/or real shells.

   d. If you have freshwater snails, you may want to add a few to the aquarium. Alternatively, these can be used at a separate water-snail station and added to the aquarium at the end of the session.

8. Update the **Time Travel Terrarium.**

   a. Tape **Terrarium Background—Time Period #2** (page 136B) to the back of the terrarium, using a minimum of tape.

b. Remove the lava rock or dirt from Session 2 and replace it with regular soil. (You can choose to add a layer of soil on top of the lava rock, but this will make it more difficult to use if you teach the unit again in the future.)

c. Remove the vials, cups, lights, and other "volcanic" material from Session 2.

d. Plant any examples of early plants you've acquired. (If you just have cuttings, these can be "planted" to make them look as if they're growing.)

e. Add the plastic scorpion(s), plastic millipede(s), and live land snail(s), and any other live or plastic organisms appropriate to the time period.

9. Copy and set aside the **Major Evolutionary Events—Time Period #2** sheet (page 127), ready to post over the **Class Time Line.**

On the Day of the Activity

1. Make copies of the station sheets and set up the remaining core stations.

   a. Set out the two copies of **Early Land Plants** (page 120) at a station.

   b. Set out the two copies of **Arthropods** (page 121) at a station.

   c. Set out the two copies of **Fossils—Time Period #2** (page 128) at a station.

   d. Set out two copies of **Continental Drift—Time Period #2** (page 129) at a station.

2. Place the organisms in their containers and set up the additional stations.

3. Gather the students' **Time Travel Journals** and **Organism Keys** from the previous session, ready to distribute. Have the appropriate **Fossils—Teacher's Answer Sheet** for this session (page 65) readily available.

4. Put out sets of **journal pages** labeled **Time Period #2** for students to add to their journals.

5. Set aside the third copy of the **Continental Drift—Time Period #2** sheet to add to the **Life through Time** wall chart.

6. Copy **The Age of** _____ sign (page 78) or write the information on a blank 8 ½" x 11" sheet of paper.

7. Set out the overhead transparencies **From Worm to Insect, Algae Reproduction,** and **Other Early Invertebrates** from previous sessions (pages 25, 71, and 75).

8. Cover the aquarium and terrarium until the unveiling.

9. Set aside the adding-machine-tape strip "410 MYA–286 MYA," ready to add to the **Class Time Line.**

■ Preparing for Time Travel

1. Welcome students to the second day of time travel! Tell them the time period they'll be traveling to today is 544 million–410 million (544,000,000–410,000,000) years ago, or 544–410 MYA. Write it on the board or point it out on the **Class Time Line** you added to in Session 2. This is a period of 134 million (134,000,000) years.

2. Distribute the **Time Travel Journals,** the sets of new journal pages for this time period, and the **Organism Keys.** As you're handing these out, ask your students to review what they think the Earth will be like during this time period, including land, water, plants, animals, weather, atmosphere, and continents. They can refer back to their predictions from the end of the previous session, or review the last session's major evolutionary events posted over the **Class Time Line.**

3. Briefly discuss some of the students' predictions. Accept all answers.

4. Tell students that as with the last session, they're going to investigate an age from long ago. Activities have again been set up at stations around the room to help them understand what life and the environment were like during this period. Explain that their job once again will be to visit each station, record their observations, and answer questions in their journals.

5. Briefly go over each station's activities by pointing out the station materials and providing the overviews that follow.

■ Stations Overview

Stations 1 and 2: Time Travel Aquarium and Terrarium stations

Remind students to observe the new dioramas, make a drawing of each, and label the different organisms in their drawings. Remind them to include organisms represented by plastic figures and illustrations, and to circle any organisms they identify on their **Organism Key.**

Station 3: Continental Drift

Remind students to draw the continents as they appeared during this time period, and to read the weather and atmosphere report.

Station 4: Fossils

Remind students to examine the fossil pictures at the station, try to guess what organisms they once were, use their **Organism Key** to identify any they can, and write them down in their journals.

Stations 5 and 6: Early Land Plants and Arthropods

Tell the students that at these stations they'll turn to their journals and look at the pictures on the Arthropods and Early Land Plant sheets at the stations. They should notice what the *arthropods* have in common and what the *early land plants* have in common. They'll then write two or more statements about each in the designated spots in their journals.

Additional stations 7 through 10

For the organism adaptations stations, students should once again note as much as they can about these organisms—what they might eat, how they move around, what features they have for self-protection, and what their predators might be. If you used the trilobite fossil or shed exoskeleton, students will need to act like paleontologists, looking for evidence that answers the same questions as for the living organisms.

■ Traveling through the Stations

As in Session 2, divide students into teams of two, assign them to their first stations, then let them rotate through all stations in any order at their own pace. Remind them to avoid stations that are crowded and return to them later. Teams that finish all the stations early may go back to station(s) of their choice to observe further.

■ Time Travel Debrief

1. Focus students on the **Life through Time** wall chart for a review of the stations. They'll need their **Time Travel Journals** and **Organism Keys** for the discussion.

2. Remove the background illustrations from the aquarium and terrarium and place them on the **Life through Time** wall chart.

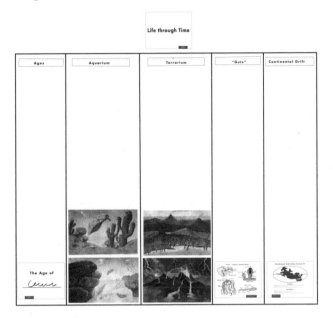

3. Debrief the **Aquarium, Terrarium,** and **Fossils** stations as in Session 2.

4. Debrief the **Early Land Plants** station.

   a. Place the overhead transparency **Early Land Plants** on the projector. Ask a few students to share their journal statements about early land plants. Ask them how they think these early plants felt to the touch, where they lived, and how they reproduced.

   b. If students didn't guess how the first land plants reproduced, explain that when wet, these plants expelled their reproductive cells and let them meet up in the water with reproductive cells from other plants. Remind students that this is also how early plantlike organisms such as algae reproduced, and point out that this similarity is another clue that these early true plants were related to algae.

   c. Ask students how they think this reproductive strategy affected where the first land plants could and could not grow. [It doesn't work in dry areas, so the first land plants grew only in moist regions. The dry areas of land were still lifeless.]

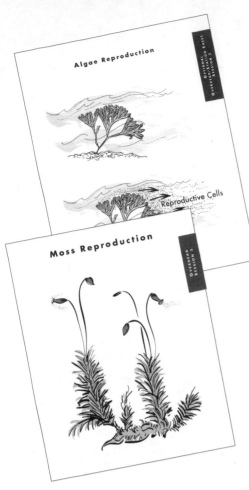

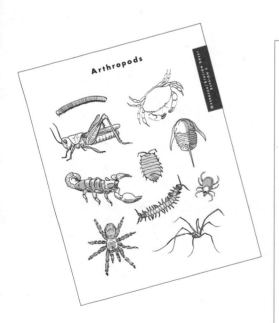

d. Alternate the overhead transparency **Algae Reproduction** from Session 2 with the **Moss Reproduction** transparency for this session. Tell students that scientists think that mosses were one of the first land plants.

e. Ask students how they think plants changed (adapted) to live on land. Point out the stiff stalk or stem they developed, in order to stand upright.

f. Ask students why they think it was important for land plants to stand erect. [Plants need sunlight to photosynthesize. Erect stems allowed early land plants to grow as high as—or above—their neighbors; early competition for sunlight may have led to stiffer, erect stems.]

g. Point out that although the club mosses and horsetails of today are small, during the time period we're visiting—with pervasively humid conditions—they were huge, and there were giant forests of them.

5. Debrief the **Arthropods** station.

a. Put the overhead transparency **Arthropods** on the overhead projector. Ask your students to turn to their journals, and ask a few students to share their statements about arthropods. Be sure the following points are made:

Arthropods

- **They appeared in huge numbers and with great diversity during this period.**

- **They had exoskeletons** (a hard outer shell). This was a breakthrough, as all earlier animals were soft. (At this evolutionary stage, no life-form yet had bones inside the body.) In order to grow, they needed periodically to shed (molt) their exoskeletons.

- **Arthropods had segmented body parts.** Like the earthworms that appeared in the previous period, arthropod bodies are divided into a linear series of similar parts—but with different patterns in different

---

species. For example the insect had head, thorax, and abdomen, while the spider had cephalothorax (head and thorax combined) and abdomen.

- **Arthropods had jointed appendages** modified to form antennae, mouthparts, and reproductive organs. Again, the patterns were very different in different species.

b. Ask students how they think having a hard outer skeleton (exoskeleton) affects how big the arthropods can get. [It limits their size. The exoskeleton is hard but not as strong as bone; it can't support the weight of great mass.]

c. Show students the overhead transparency **From Worm to Insect** from Session 1. Remind them that it's an artist's depiction of how segmented worms may have evolved into insects over millions of years. Point out the many different combinations of mouthparts, body segments, and legs that different arthropods have evolved from a common ancestor.

d. Ask a few students to compare their recent statements about arthropods with those they wrote in Session 2 about other early invertebrates. During discussion, alternate the overhead transparency **Other Early Invertebrates** from the previous session with the **Arthropods** transparency. Mention the following points, if students don't:

From Worm to Insect

What Arthropods and Earlier Invertebrates Have in Common

- **They have no endoskeletons** (bones inside body).

- **They have segmented bodies.**

- **They have very similar nervous systems** constructed along the same basic plan: a brain at the back of the head leading to a series of nerve bundles (ganglia) running down the back.

6. Debrief the **Organism Adaptations** stations.

Ask students to refer to the Organism Adaptations pages in their journal. Revisit each animal and plant from the stations, asking the students for a few of their ideas on each organism's adaptations. If possible, do this by gathering around the organism being discussed, or by placing it in front of the class.

7. Debrief the **Continental Drift** station.

a. Review how the continents appeared during the previous time period on the **Life through Time** wall chart. Ask students how the continents drifted during the time period they just studied. If not mentioned, point out the following:

- As the period begins, land has formed into two large continents and other fragments around the globe.
- Almost all land lies between the tropics and the equator, and many glaciers are melting.
- Melting ice sheets create huge, warm, shallow seas that provide excellent conditions for early sea life to flourish.
- What will become Africa, South America, Australia, and India have moved south of the equator.
- By the end of this period, the land has collected into one huge continent plus several smaller ones.

b. Add the station sheet **Continental Drift—Time Period #2** to the **Life through Time** wall chart under the heading "Continental Drift."

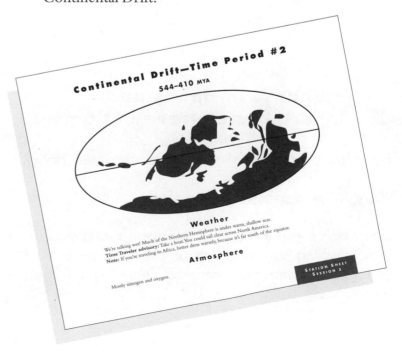

■ Major Evolutionary Events

1. Ask students to consider all they've just discussed, then brainstorm some major evolutionary changes that took place in the organisms or their habitats during the period they just studied, referring to the **Major Evolutionary Events—Time Period #2** list as a content checklist.

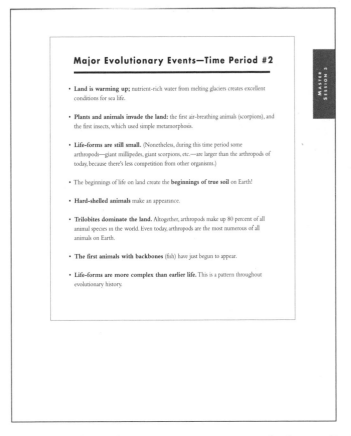

Major Evolutionary Events—Time Period #2

- **Land is warming up;** nutrient-rich water from melting glaciers creates excellent conditions for sea life.

- **Plants and animals invade the land:** the first air-breathing animals (scorpions), and the first insects, which used simple metamorphosis.

- **Life-forms are still small.** (Nonetheless, during this time period some arthropods—giant millipedes, giant scorpions, etc.—are larger than the arthropods of today, because there's less competition from other organisms.)

- The beginnings of life on land create the **beginnings of true soil** on Earth!

- **Hard-shelled animals** make an appearance.

- **Trilobites dominate the land.** Altogether, arthropods make up 80 percent of all animal species in the world. Even today, arthropods are the most numerous of all animals on Earth.

- **The first animals with backbones** (fish) have just begun to appear.

- **Life-forms are more complex than earlier life.** This is a pattern throughout evolutionary history.

2. Ask students how their observations compare with the predictions they made in Session 2.

3. Post the sheet **Major Evolutionary Events—Time Period #2** above this session's strip of the **Class Time Line.**

■ Most Representative Organism of the Age

1. Tell students that, as in the previous session, the class will vote for a name for the time period they just studied. Remind them that they'll name it for what they consider to have been the most representative organism; the one that best represents the period. They'll vote after all the candidates have made their speeches.

2. Hand out the cut-up speeches to those who'll play each role.

3. Have the volunteers read their speeches, in numerical order. Remember to introduce each candidate and write its name on the board. After the speeches, you may again want to ask if anyone has another organism to nominate and make an improvised campaign speech for.

4. Reconvene the class and hold the election. Remind students to vote only once, and to vote for the organism they think best represents the time period. Review the candidates with the class:

- The Age of Sea Squirts
- The Age of Trilobites
- The Age of Sea Scorpions
- The Age of Liverworts

Let students know they may also vote for a name that would include many candidates:

- The Age of Invertebrates (Animals without Backbones)

5. Conduct and tally the vote. Write the name of the winner on the **Age of_____** sign, a blank sheet, or directly on the **Life through Time** wall chart. Have students write it on the aquarium and terrarium drawing page of their **Time Travel Journals.** If you made a sign, add it to the wall chart, under the "Ages" heading.

■ Class Time Line

1. Attach to the **Class Time Line** the strip of adding-machine tape you pre-cut for the next session, "410 MYA–286 MYA." Tell your students the next session will focus on this period.

2. Write the years of the next time period on the board using any or all of the following forms:

- 410 million–286 million years ago
- 410,000,000–286,000,000 years ago
- 410–286 MYA

3. Have students write in their **Time Travel Journals** their predictions about the organisms and habitat changes they think they'll encounter in the next time period.

4. Ask students to bring in organisms you'll need for the next session, to be used at the organism adaptation stations.

5. If there's time, you may wish to have one or more students place this session's **Tree of Life Organism Cards** on the **Tree of Life** wall chart as shown. Alternatively, you can do this yourself before the next session.

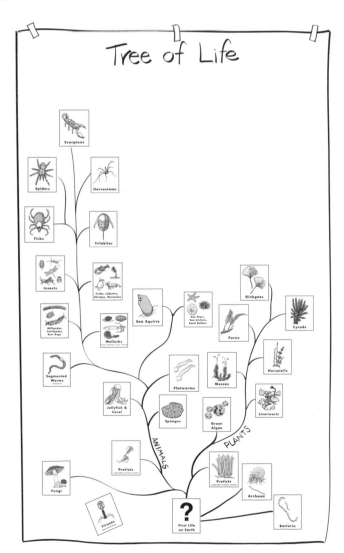

■ Going Further

1. Incorporating Related Activities

Any activities you have from other curricula relating to soil, early land plants, shellfish, insects, spiders, or other invertebrates would be excellent additions at this point.

2. Songs

Two songs at the back of the guide make good connections to this session: the funk song about spiders, *Arachnidae* (page 322), and the bilingual (Spanish/English) salsa song, *Dancing With The Insects (Bailando Con Los Insectos)* (pages 323–324). You can preview these songs at www.moo-boing.com.

Script for Most Representative Organism Election for
Time Period #2: 544–410 MYA

- -

1

Sea Squirt: *[Turn head to one side and suck in air, turn head to the other side and blow it out. Continue this motion occasionally throughout the campaign speech. Speak in an old, great-grandparent type voice.]* Hey, how ya doin'? I'm a sea squirt. You know what us sea squirts did and still do? We suck water in and squirt it out. That's about it–but we like it! You young whippersnappers need video games and other stuff, but not us! We suck water in and squirt it out—and we like it! Why are we important? We're the first animal with the beginnings of a backbone. All other animals with backbones—fish, amphibians, reptiles, birds, and mammals—should be grateful. You're not, but you should be! Think of where you'd be without a backbone. 'Course, it's only in our young larval stage that we have this, and not as adults. You know what we do as adults? We suck water in and squirt it out—and we like it! This should be called the "Age of Sea Squirts"!

- -

2

Trilobite: We're trilobites, and we ruled! Sea squirts may be the first animals with the beginnings of a backbone, but that's **all** they are. Sure, animals with backbones that were more advanced than the sea squirt, like fish and birds, ruled other time periods, but not this one! All sea squirts did was squirt sea. See, ya little squirt, us trilobites, on the other hand, we were crawling and swimming everywhere. Look at that picture up there. *[Points to background illustration from diorama.]* Tons of different kinds of trilobites all throughout the water, and even onto the beaches. If you look at fossils from this time period, it's trilobites that you're mostly going to find. Eventually we all died off, but during this time period, we ruled the waters. So you need to call this the "Age of Trilobites"!

- -

3

Sea Scorpion: *[Slow, low and creepy voice.]* Yeah, there may have been a lot of trilobites, but I was the terror of the deep, and I ate trilobites for breakfast. *[Look at trilobite hungrily.]* Lunch and dinner too, and maybe a midnight snack. I was a seven-foot-long invertebrate sea monster, and you wouldn't want to meet me underwater. Yeah, that's right, **seven feet long!** You're lucky we're extinct now, but in our day we were awesome. You wouldn't dare call this anything but the "Age of Sea Scorpions," would you???

4

Liverwort: *[Looking at trilobites and sea scorpions.]* You both went extinct, but I—the liverwort—am still around. So my name isn't exactly beautiful, but I was the first land plant. Without land plants, there wouldn't be land animals! It wasn't easy either. In the water we had water to hold us up, keep us moist, and spread our spores around to reproduce. On land we had to use whatever water we could get, and evolve a stalk to hold us up. "Wort" doesn't mean "plant" for nothing; this period should be named in our honor—the "Age of Liverworts"!

Moss Reproduction

Early Land Plants

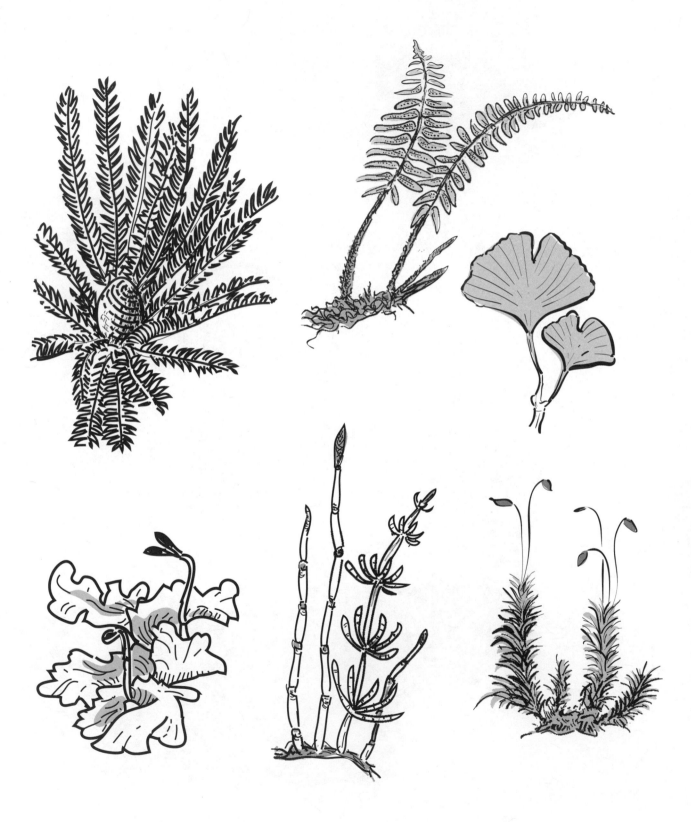

Arthropods

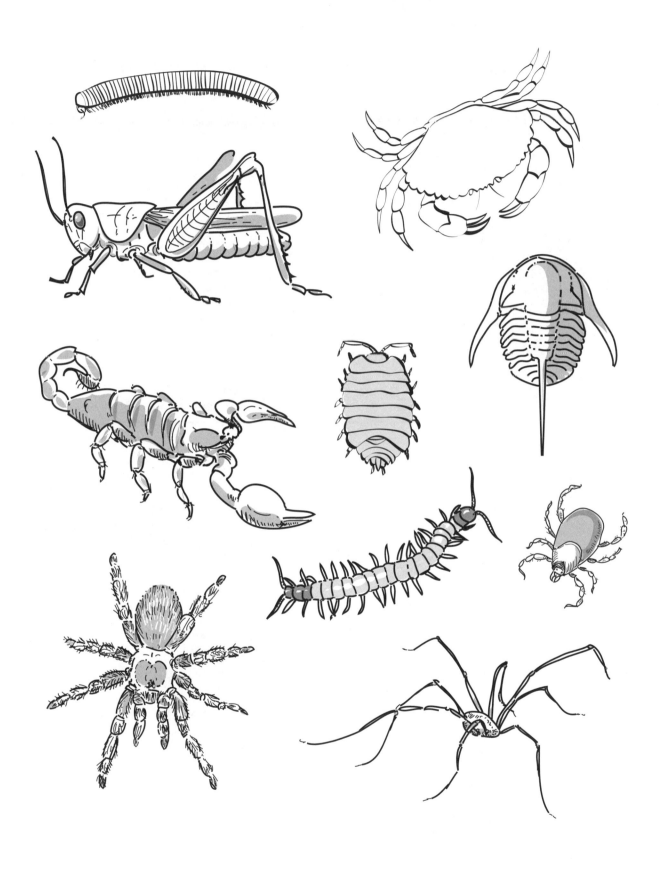

Liverworts

Ferns

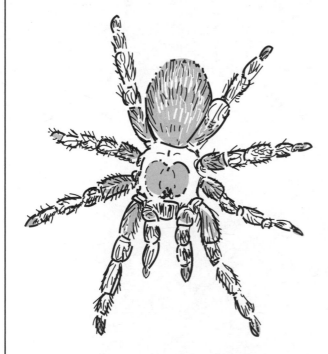

Spiders

Cycads

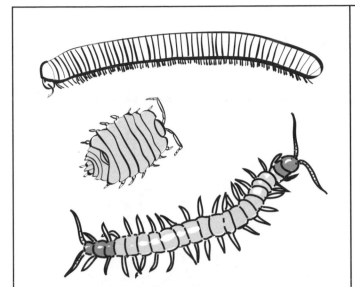

Millipedes, Centipedes, Sow Bugs

Scorpions

Crabs, Lobsters, Shrimps, Barnacles

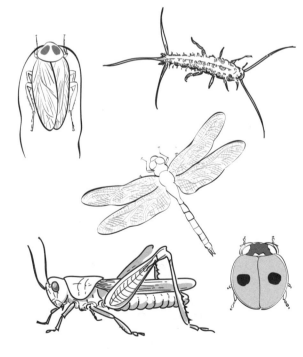

Insects

Tree of Life Organism Cards
Time Period #2

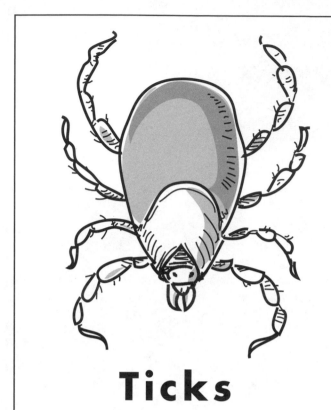

Ticks

Mosses

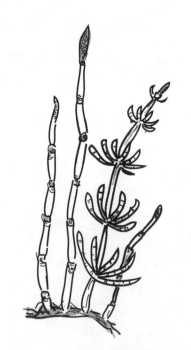

Horsetails

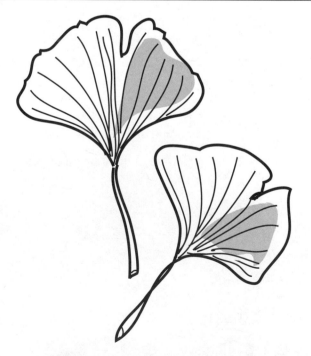

Ginkgoes

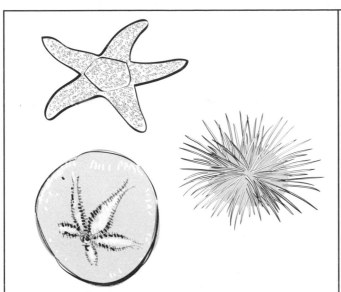

Sea Stars, Sea Urchins, Sand Dollars

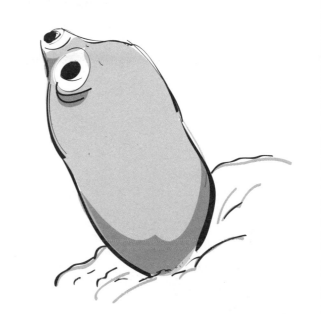

Sea Squirts

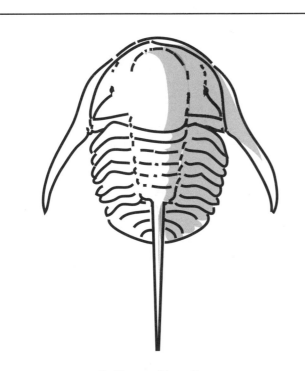

Trilobites

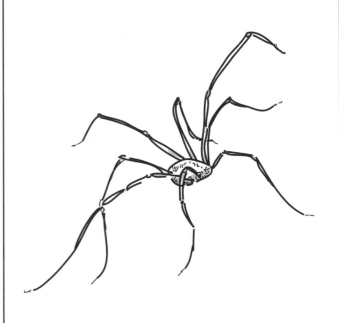

Harvestmen

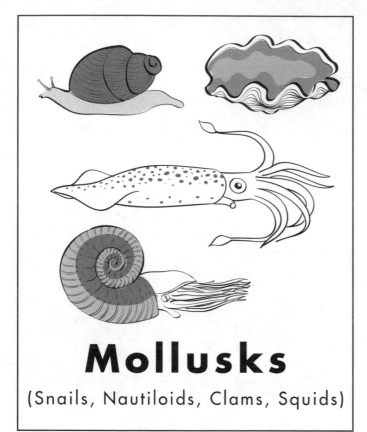

Mollusks
(Snails, Nautiloids, Clams, Squids)

Major Evolutionary Events—Time Period #2

- **Land is warming up;** nutrient-rich water from melting glaciers creates excellent conditions for sea life.

- **Plants and animals invade the land:** the first air-breathing animals (scorpions), and the first insects, which used simple metamorphosis.

- **Life-forms are still small.** (Nonetheless, during this time period some arthropods—giant millipedes, giant scorpions, etc.—are larger than the arthropods of today, because there's less competition from other organisms.)

- The beginnings of life on land create the **beginnings of true soil** on Earth!

- **Hard-shelled animals** make an appearance.

- **Trilobites dominate the land.** Altogether, arthropods make up 80 percent of all animal species in the world. Even today, arthropods are the most numerous of all animals on Earth.

- **The first animals with backbones** (fish) have just begun to appear.

- **Life-forms are more complex than earlier life.** This is a pattern throughout evolutionary history.

Fossils—Time Period #2

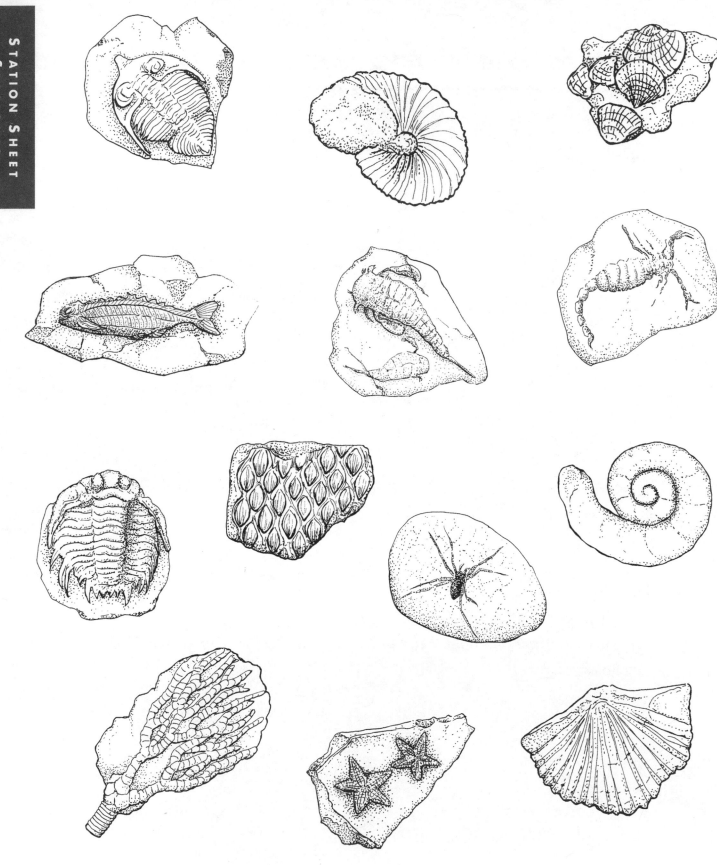

Continental Drift—Time Period #2

544–410 MYA

Weather

We're talking wet! Much of the Northern Hemisphere is under warm, shallow seas.
Time Traveler advisory: Take a boat. You could sail clear across North America.
Note: If you're traveling to Africa, better dress warmly, because it's far south of the equator.

Atmosphere

Mostly nitrogen and oxygen.

Time Travel Journal —Time Period #2

544 million to 410 million years ago

The Age of _____

Draw and label animals and plants you identify.

Time Travel Aquarium

Time Travel Terrarium

Fossils

Label any fossils you're able to identify.

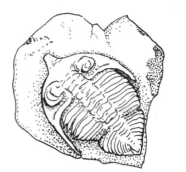

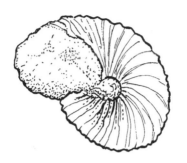

_____

_____

_____

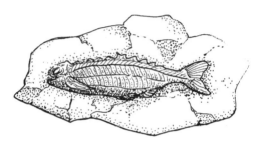

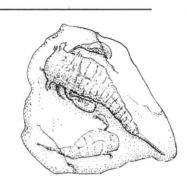

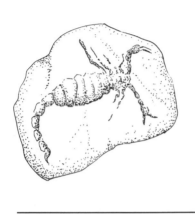

_____

_____

_____

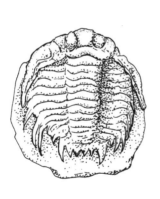

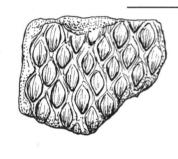

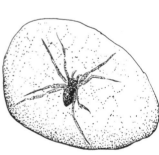

_____

_____

_____

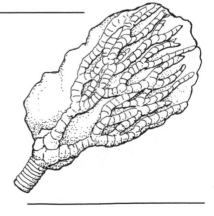

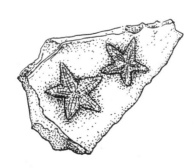

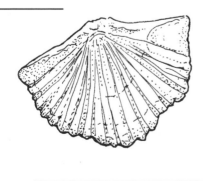

_____

_____

_____

Early Land Plants

Look at the pictures and write two or more statements about what these organisms have in common and how they are different.

_____

_____

_____

_____

_____

STUDENT HANDOUT SESSION 3

Arthropods

Look at the pictures and write two or more statements about what these organisms have in common and how they are different.

_____

_____

_____

_____

_____

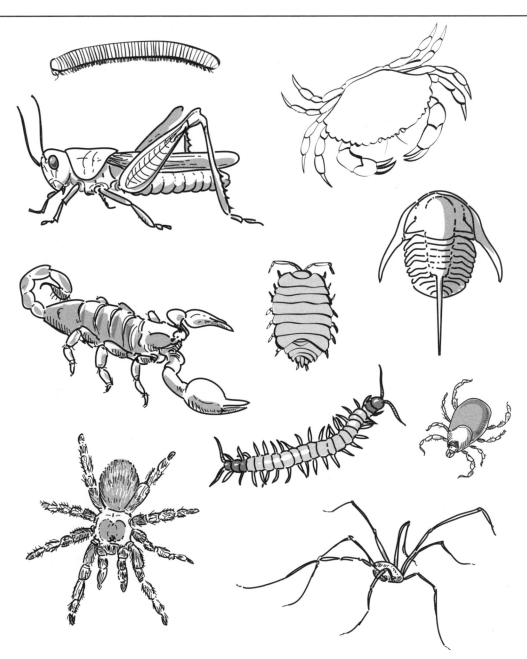

Organism Adaptations

Carefully observe the animals and plants, and write your best guesses to these questions:

Animal	What does it eat?	What eats it?	How does it move around?	How does it protect itself?

Plant	How does it feel?	What kind of environment does it live in? (Wet? Dry? Very dry?)	How does it reproduce?

Continental Drift

Draw the continents as they appeared in this time period, referring to the station sheet. Describe the weather and atmosphere.

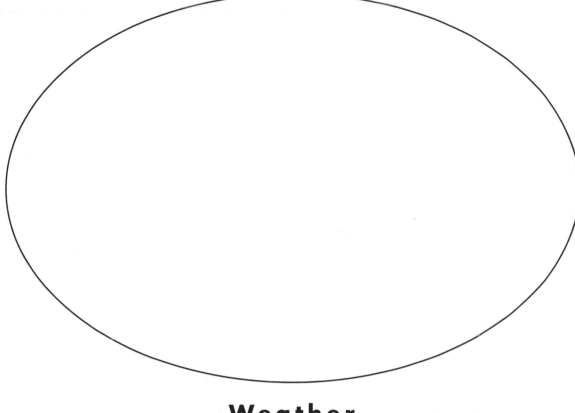

Weather

_____

_____

_____

_____

Atmosphere

_____

_____

_____

_____

Predictions for Next Time Period

Write your predictions for how organisms, land, water, weather, and atmosphere may change in the next time period.

_____

_____

_____

_____

_____

_____

_____

_____

_____

_____

_____

_____

_____

_____

_____

_____

_____

Aquarium Background —Time Period #2 • *Session 3*

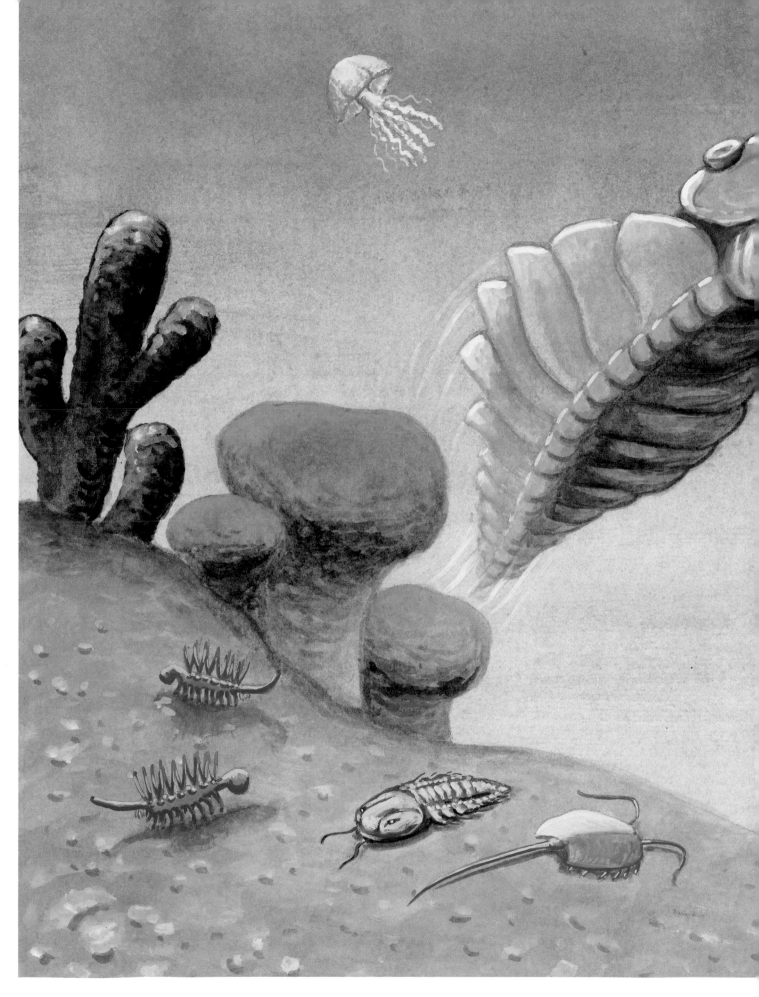

Terrarium Background —Time Period #2 • *Session 3*

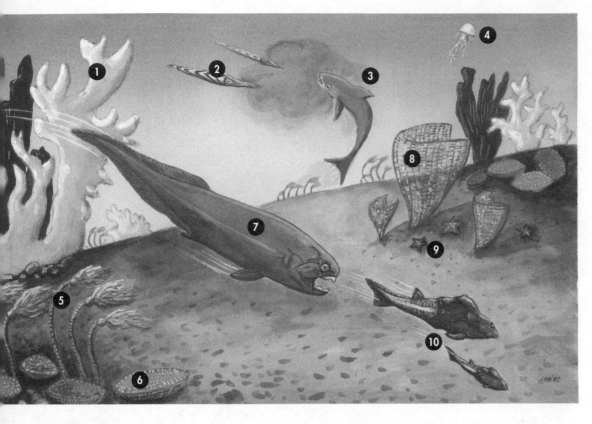

Aquarium

1. Corals
2. Nautiloids
3. Early Shark
4. Jelly
5. Crinoids
6. Flat Corals
7. Placoderm (Early Jawed Fish)
8. Bryozoans
9. Seastars
10. Acanthodians (Early Jawed Fish)

Terrarium

1. Giant Club Mosses
2. *Meganeura* (Giant Dragonflies)
3. Giant Cockroach
4. *Sigillaria*
5. *Diplocaulus*
6. *Ichthyostega*
7. Horsetails

Overview

This session explores life on Earth during the period 410–286 MYA. As this age begins, fish are abundant in the seas and amphibians have invaded the land, where the first forests are now growing. Cone-bearing trees, with their water-retaining seeds and water-conserving needles, have been able to move into drier areas than earlier species could. Toward the end of this period, the first reptiles arrive on the scene.

> This session represents the Devonian and Carboniferous Periods of the Paleozoic ("Ancient Life") Era of the Phanerozoic ("Evident Life") Eon.

As before, the students rotate through a series of stations, including the Time Travel Aquarium and Terrarium, organisms that represent the time period, Continental Drift, Fossils, Amphibians, Fish, and Conifers. They use their **Time Travel Journals** and **Organism Keys** to record observations and guess at adaptations, and to identify the organisms they see.

The wrap-up discussion centers around fish and amphibians. The **Life through Time** wall chart continues to grow, the period's most significant evolutionary events are reviewed and placed on the **Class Time Line,** and the organisms featured during the session are added to the **Tree of Life** wall chart. The most representative organism of this new time period is elected after candidates' speeches. Once again, students make predictions for what will occur in the next time period.

■ What You Need

For the teacher:
- ❑ the appropriate **Fossils—Teacher's Answer Sheet** for this session (page 66), from your set

For the class:
- ❑ 1 copy of this session's **Most Representative Organism Script** (pages 152–153)
- ❑ 1 overhead transparency of **Conifer Reproduction** (page 154)
- ❑ 1 overhead transparency of **Fish** (page 155)
- ❑ 1 overhead transparency of **Amphibians** (page 156)
- ❑ 1 overhead transparency of **Conifers** (page 157)
- ❑ the overhead transparencies **Algae Reproduction** and **Moss Reproduction** from previous sessions (pages 71 and 119)
- ❑ 1 copy of this session's **Tree of Life Organism Cards** (pages 158–159) to add to the **Tree of Life** wall chart

❑ 1 copy of **The Age of** _____ sign (page 78)
(Remember, you may instead write this information on a blank
8 ½" x 11" sheet of paper or directly on the **Life through Time**
wall chart after the election, to simplify.)

❑ 1 copy of **Major Evolutionary Events—Time Period #3**
(page 160)

❑ the adding-machine-tape strip you cut for the next session,
"286 MYA–65 MYA," to add to the **Class Time Line**

❑ the **Time Travel Aquarium** and **Time Travel Terrarium**
from the previous session

❑ the two large cloths from the previous session, to cover the
aquarium and terrarium

❑ an overhead projector and screen

For the Time Travel Aquarium station:

❑ 1 copy of **Aquarium Background—Time Period #3** (page
160A)

❑ enough dechlorinated water to top off the aquarium, if needed

❑ artificial coral (Do not collect coral from the wild.)

❑ plastic sharks

❑ plastic bony fishes

❑ goldfish, mosquito fish, or whatever kind of live fish you have
(Small fish can be caught in a local pond with a net. We do *not*
recommend putting your student's pet fish in the Time Travel
Aquarium. Please see note on page 21 about mosquito fish.)

For the Time Travel Terrarium station:

❑ 1 copy of **Terrarium Background—Time Period #3** (page
160B)

❑ one or more of these early plants:
___ horsetail
___ small ferns
___ ginkgo (Popular urban ornamental tree in people's yards or
on West Coast sidewalks.)
___ cycad (Early plant resembling—but different from!—a palm
or tree-fern.)
___ cones and needles, or just cones, from any type of cone-
bearing tree (pine, redwood, fir, etc.) Ideally, get both male
cones (they're the small ones) and female cones (larger) from
a single type of tree.

❑ plastic dragonfly

❑ plastic salamander

For the remaining core stations:

❑ 2 copies of the station sheet **Fish** (page 155)

❑ 2 copies of the station sheet **Amphibians** (page 156)

- ❑ 2 copies of the station sheet **Conifers** (page 157)
- ❑ 2 copies of the station sheet **Fossils—Time Period #3** (page 161)
- ❑ 3 copies of the station sheet **Continental Drift—Time Period #3;** two for the station and one for the **Life through Time** wall chart (page 162)

These are the seven core stations for this session:

1) Time Travel Aquarium	5) Fish
2) Time Travel Terrarium	6) Amphibians
3) Continental Drift	7) Conifers
4) Fossils	

In addition to these seven, you'll need at least three organism adaptations stations from the following list.

For the additional three or more stations:
- ❑ several containers to "house" organisms at the stations (Use whatever containers seem appropriate for the organism.)
- ❑ fish (goldfish, tetra, guppies, etc.)
- ❑ horsetail
- ❑ fern
- ❑ salamander★
- ❑ frog★
- ❑ cockroach
- ❑ fish skeleton
- ❑ dead fish, whole or partly dissected
- ❑ amphibian skeleton

★ **Remember to use care when considering these organisms. See note on page 21.**

For each student:
- ❑ **Time Travel Journal** from the previous session
- ❑ **Organism Key** from the previous session
- ❑ 1 set of **Time Travel Journal** pages labeled **Time Period #3** (pages 163–170; 8 pages total) to add to journal *(Don't copy these until you've added this session's organisms to the **Organism Adaptations** page; see "Getting Ready," below.)*

■ Getting Ready

Before the Day of the Activity

1. Copy and cut up this session's **Most Representative Organism Script** (pages 152–153) so each character's part is on a separate sheet. If other adults are reading the parts, contact them to schedule their time.

2. Write the names of the organisms you've acquired for this session in the left-hand column on the **Organism Adaptations** page of the students' journals.

3. Copy the **Time Travel Journal** pages marked **Time Period #3** (pages 163–170; 8 pages total), punching holes if needed.

4. Copy and set aside the overhead transparencies for this session: **Conifer Reproduction, Fish, Amphibians,** and **Conifers** (pages 154–157).

5. Label the containers you'll be using for the organisms at the **Organism Adaptations** stations. They should first be labeled "Fish" or "Conifers," then more specifically by kind, such as "goldfish" or "pine."

6. Copy and cut up this session's **Tree of Life Organism Cards** (pages 158–159), ready to place on the **Tree of Life** wall chart.

7. Update the **Time Travel Aquarium.**

   a. Top off the water level if needed.

   b. Lightly tape **Aquarium Background—Time Period #3** (page 160A) to the back of the aquarium.

   c. Add the artificial coral, plastic sharks, plastic bony fishes, and whatever kind of live fish you have.

Note: One or two examples of each organism should be left out of the aquarium to be studied at organism adaptation stations. They can later be added to the aquarium.

8. Update the **Time Travel Terrarium.**

   a. Lightly tape **Terrarium Background—Time Period #3** (page 160B) to the back of the terrarium.

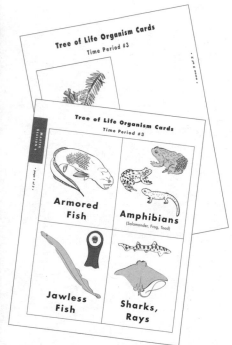

b. Plant any horsetails, small ferns, ginkgoes, and/or cycads you've acquired.

c. Add cones or needles from any type of cone-bearing tree.

d. Add the plastic dragonfly and plastic salamanders.

9. Copy and set aside the **Major Evolutionary Events—Time Period #3** sheet (page 160), ready to post over the **Class Time Line.**

On the Day of the Activity

1. Make copies of the station sheets and set up the remaining core stations.

   a. Set out the two copies of **Fish** (page 155) at a station.

   b. Set out the two copies of **Amphibians** (page 156) at a station.

   c. Set out the two copies of **Conifers** (page 157) at a station.

   d. Set out the two copies of **Fossils—Time Period #3** (page 161) at a station.

   e. Set out two copies of **Continental Drift—Time Period #3** (page 162) at a station.

2. Place the organisms in their containers and set up the additional stations.

3. Gather the students' **Time Travel Journals** and **Organism Keys** from the previous session, ready to distribute. Have the appropriate **Fossils—Teacher's Answer Sheet** for this session (page 66) readily available.

4. Put out sets of **journal pages** labeled **Time Period #3** for students to add to their journals.

5. Set aside the third copy of the **Continental Drift—Time Period #3** sheet to add to the **Life through Time** wall chart.

6. Copy **The Age of** _____ sign (page 78) or write the information on a blank 8 ½" x 11" sheet of paper.

7. Set out the overhead transparencies **Algae Reproduction** and **Moss Reproduction** from previous sessions (pages 71 and 119).

8. Cover the aquarium and terrarium until the unveiling.

9. Set aside the strip "286 MYA–65 MYA," ready to add to the **Class Time Line.**

■ Prepare for and Begin the Time Travel

1. Welcome your students to the time period they'll be traveling to today: 410 million–286 million (410,000,000–286,000,000) years ago, or 410–286 MYA. Write it on the board or point it out on the **Class Time Line** you added to in Session 3. This is a period of 124 million (124,000,000) years.

4.5 BYA—544 MYA	544 MYA—410 MYA	410 MYA—286 MYA

2. Distribute the **Time Travel Journals,** the sets of new journal pages for this time period, and the **Organism Keys.** Ask students to review their predictions for what the Earth will be like during this time period, including land, water, plants, animals, weather, atmosphere, and continents.

3. Briefly discuss some of the students' predictions. Accept all answers.

4. Tell students that, as in previous sessions, they'll get a chance to investigate an age from long ago. Activities have again been set up at stations around the room to help them understand what life and the environment were like during this period. Once again they'll visit each station, record their observations, and answer questions in their journals.

5. If you feel your students need a refresher, review the standard stations (aquarium, terrarium, continental drift, fossils, and organism adaptations) with them and then provide a brief overview of the new stations, pointing out the station materials.

6. Divide students into teams of two, assign each team to its first station, and let them rotate through the others at their own pace and in any order.

**Stations recap—
Session 4/Time Period #3**

1) Time Travel Aquarium
2) Time Travel Terrarium
3) Continental Drift
4) Fossils
5) Fish
6) Amphibians
7) Conifers
8) through 10)—whatever additional stations you've chosen

■ Time Travel Debrief

1. Focus the students on the **Life through Time** wall chart. Be sure they have their **Time Travel Journals** and **Organism Keys** for the discussion.

2. Remove the background illustrations from the aquarium and terrarium and place them on the **Life through Time** wall chart.

3. Debrief the **Aquarium, Terrarium,** and **Fossils** stations as in Session 2.

4. Debrief the **Fish** station.

Put the **Fish** transparency on the projector, and ask a few students to share their statements about fish. Be sure the following points come up:

The first beginnings of a skeleton have been found in the larval sea squirt, but fish were the first to have them their whole lives.

Fish

- **Fish have backbones (some of cartilage, some of bone).** Earlier animals were soft, or had only exoskeletons. The evolution of the endoskeleton (bones on the inside) allowed animals to grow larger and faster. Ask students why they think that is. [With no external bones to contain their growth, animals could expand—not without limits, but more than earlier life-forms "hemmed in" by exoskeletons. An internal skeleton also provided better structural support for the limbs, backbone, and body cavities.]

- **Fish have teeth;** not for chewing, but for grabbing food.

- **The skin of fish are scaly or plated.**

- **Fish eggs are jelly-like** and, though laid in large quantities, they're generally **laid individually, not in a continuous string.**

- **Fish evolved as more complex than earlier life-forms,** following the classic evolutionary pattern.

5. Debrief the **Amphibians** station.

    a. Put the overhead transparency **Amphibians** on the projector. Ask a few students to share their statements about amphibians. Bring out the following, if your students don't:

Amphibians

- **Their backbones are made of bone, not cartilage.**

- **Like fish, they have teeth;** not for chewing, but for grabbing food.

- **They have a muscular tongue** to move food around or to catch prey.

- **Amphibians have smooth, water-permeable skin** (skin that lets moisture pass through), through which they take in oxygen from the water. Because they can also *lose* moisture through their skins, most amphibians are unable to live in dry areas (although toads, for example, do fine by brooding their young in their mouths or even in pockets on their backs!).

- **They are freshwater dwellers,** and can't live in salt water.

- **Amphibian eggs, like fish eggs, are jelly-like—but they lay them in continuous strings,** which helps protect them from drying out. Even so, amphibian eggs can't survive in dry areas.

- **Amphibian evolved legs that stick out from the sides of their bodies;** they have trouble lifting themselves off the ground.

- **Amphibians evolved the first eyelids and eardrums.** Eyelids protect the eye and keep it moist outside of water. Eardrums may also have been adaptations for living out of water, protecting the inner ear from damage and helping to channel and focus sound.

---

b. Ask a few students to share their statements about what the organisms from the **Amphibians** sheet had in common with the organisms from the **Fish** sheet. Alternate the **Fish** and **Amphibian** transparencies on the overhead projector during the discussion. Bring up the following if your students don't:

What Fish and Amphibians Have in Common

- **A backbone.** Their skeletons match up, bone type for bone type. They both have pelvis bones, ear bones, arm bones, etc.

- **Gills.** Amphibians use gills during some portions of their lives.

- **Teeth,** not for chewing, but for grabbing food.

- **Jelly-like eggs.**

- **"One-way traffic" guts.**

Lobe-finned Fish

Early Amphibian

6. Debrief the **Organism Adaptations** stations.

Ask students to turn to the Organism Adaptations pages in their journal. Revisit each animal and plant from the stations, asking the students for a few of their ideas on each organism's adaptations. If possible, do this by gathering around the organism being discussed, or by placing it in front of the class.

7. Debrief the **Conifers** station.

a. Show the overhead transparency **Conifers** and ask a few students to share their thoughts on what conifers had in common.

b. Show the **Algae Reproduction** and **Moss Reproduction** transparencies from previous sessions and quickly review how these early plants reproduced.

c. Show this session's **Conifer Reproduction** transparency. If you have real cones, hold them up for your students to see.

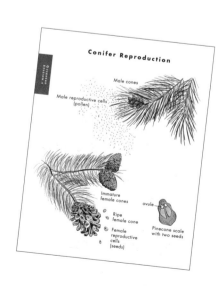

d. Ask students how they think cone-bearing trees reproduce differently from moss and algae. Explain that cone-bearing plants have small male cones that produce, then release, pollen into the wind, which carries the pollen to the larger female cones of the same or nearby trees. When the female cones become fertilized, they form watertight seeds that retain moisture. When a seed falls out of the cone and into soil, the plant embryo forms inside this moist "package," then grows out of the seed as it develops.

e. Ask students what they think the significance of these cone-bearing plants was to the adaptation to land. If they don't come up with it, help them realize that watertight seeds allow cone-bearing trees to reproduce without water, and that the thin, tough-coated needles retain moisture longer than regular leaves. Both these adaptations allowed conifers to invade drier parts of the land.

8. Debrief the **Continental Drift** station.

a. Review how the continents appeared during the previous time period on the **Life through Time** wall chart. Ask students how the continents drifted during the time period they just studied. If not mentioned, bring up the following:

- Over the course of this period, a second supercontinent formed: Laurasia. The centers of these giant land masses, Gondwana and Laurasia, were pretty stable at a time when smaller masses continued to drift.
- What will become Africa has moved to the South Pole.
- What will become North America is at the equator.
- With the weather alternating between very wet and very dry, amphibians evolve.
- By the very end of this period, Gondwana and Laurasia have merged into a single, massive supercontinent: Pangaea. It encompasses *all* land on Earth.

b. Ask what effect having all the continents connected might have had on life. [Species that were once separated by water could now migrate freely over the one continent.]

c. Add the station sheet **Continental Drift—Time Period #3** to the **Life through Time** wall chart.

■ Major Evolutionary Events

1. Ask students to consider all they've just discussed, then brainstorm some major evolutionary changes that took place in the organisms or their habitats during the period they just studied, referring to the **Major Evolutionary Events—Time Period #3** list as a content checklist.

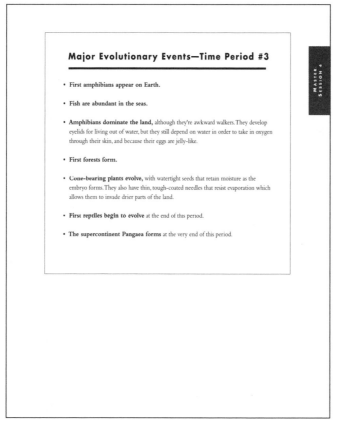

Major Evolutionary Events—Time Period #3

- First amphibians appear on Earth.

- Fish are abundant in the seas.

- **Amphibians dominate the land,** although they're awkward walkers. They develop eyelids for living out of water, but they still depend on water in order to take in oxygen through their skin, and because their eggs are jelly-like.

- **First forests form.**

- **Cone-bearing plants evolve,** with watertight seeds that retain moisture as the embryo forms. They also have thin, tough-coated needles that resist evaporation which allows them to invade drier parts of the land.

- **First reptiles begin to evolve** at the end of this period.

- **The supercontinent Pangaea forms** at the very end of this period.

MASTER SESSION 4

2. Ask students how their observations compare with the predictions they made in Session 3.

3. Post the sheet **Major Evolutionary Events—Time Period #3** above this session's strip of the **Class Time Line.**

■ Most Representative Organism of the Age

1. It's time to name this time period! Remind students they'll get to vote after each candidate has made a brief speech.

2. Hand out the cut-up speeches to those who'll play each role.

3. Have the volunteers read their speeches, in numerical order. Remember to introduce each candidate and write its name on the board.

After the speeches, ask if anyone wishes to nominate and make an improvised speech for any other organism.

4. Reconvene the class and hold the election. Remind students to vote only once, and to vote for the organism they think best represents the time period. Review the candidates with the class:

   • The Age of Conifers
   • The Age of Fish
   • The Age of Amphibians

They may also vote for a name that would include two of the candidates:

   • The Age of Fish and Amphibians, or (more challenging) the Age of First Vertebrates

5. Conduct and tally the vote. Write the name of the winner on the **Age of**_____ sign, a blank sheet, or directly on the wall chart. Have students write it in their journals. If you made a sign, add it to the chart.

■ Class Time Line

1. Attach to the **Class Time Line** the strip of adding-machine tape you pre-cut for the next session, "286 MYA–65 MYA." Tell your students the next session will focus on this period.

2. Write the years of the next time period on the board using any or all of the following forms:

- 286 million–65 million years ago
- 286,000,000–65,000,000 years ago
- 286–65 MYA

3. Have students write in their **Time Travel Journals** their predictions about the organisms and habitat changes they think they'll encounter in the next time period.

4. Ask students to bring in organisms you'll need for the next session, to be used at the organism adaptation stations.

5. If there's time, have one or more students place this session's **Tree of Life Organism Cards** on the **Tree of Life** wall chart as shown, or do this yourself before the next session.

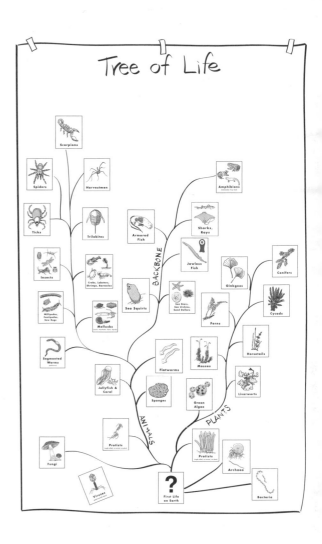

■ Going Further

1. Incorporating Related Activities
Any activities you have from other curricula relating to fish, amphibians, or cone-bearing plants would be an excellent addition at this point.

2. Songs
The *Shark's in the House* rap song at the back of this guide (page 325) is about sharks, including the giant prehistoric *Carcharodon megalodon*. *Bubba (The Frog),* on page 326, is a blues song about amphibians. Hear audio clips at www.moo-boing.com.

3. Evolve-a-Worm
Using the worm guts illustration from Session 2 as a basis, have your students imagine that a worm was put on a fictitious continent. Ask them to imagine and draw an "alien" invertebrate that evolved from it—including digestive system and other internal organs. They could continue this project each day, evolving the organism into fish, amphibian, reptile, dinosaur, bird, human, other mammal….

Script for Most Representative Organism Election for
Time Period #3: 410–286 MYA

- -

1

Redwood: The conifers rule! I should know—I'm the tallest living thing ever on Earth. Living on land can be tough on a plant. Those limp little algae couldn't take it. They need water to spread their spores around. Early land plants weren't much better. They needed water to wash around their reproductive cells. But we cone-bearing plants broke free from the wet lands and moved to dry ground. We have seeds that are watertight and carry their own moisture. Our "leaves" are needles; they're tough and don't lose much water, so we can invade drier places. This should be called the "Age of Conifers"!

- -

2

Shark: *[Puts on a huge, toothy, shark-like grin. Makes* Jaws *soundtrack noise, and a giant jaws-like motion with arms and hands.]* I've got one word for you: fish! Most life on Earth during this time was still in the waters, and guess who ruled the waters. Fish. There were armored fish, jawless fish, fish with cartilage skeletons, and fish with bones too. And then there was the baddest of all fish—sharks. We were so good at what we did that we haven't had to change much, clear up to present time! And by the way... we *still* rule the seas. Anybody care to go swimming? Vote for the "Age of Fish"!

- -

© 2003 The Regents of the University of California. May be duplicated for classroom or workshop use.

LHS GEMS • *Life through Time*

3

Amphibian: This is the dawning of the age of amphibians! Maybe modern-day salamanders and frogs are mostly pretty small, but during this time period we were HUGE! You can imagine one of us slipping out of the waters—several meters of slimy amphibian flesh. Although other early invertebrates were already living on land, we were the first animals **with** backbones to invade the land. This should be called the "Age of Amphibians."

Conifer Reproduction

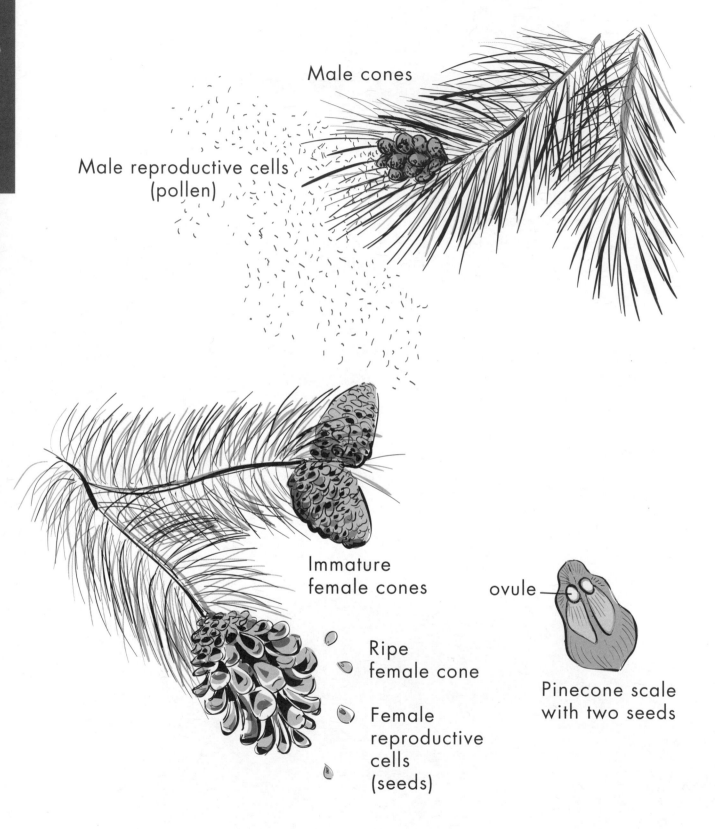

Male cones

Male reproductive cells
(pollen)

Immature
female cones

Ripe
female cone

Female
reproductive
cells
(seeds)

ovule

Pinecone scale
with two seeds

Fish

Amphibians

Conifers

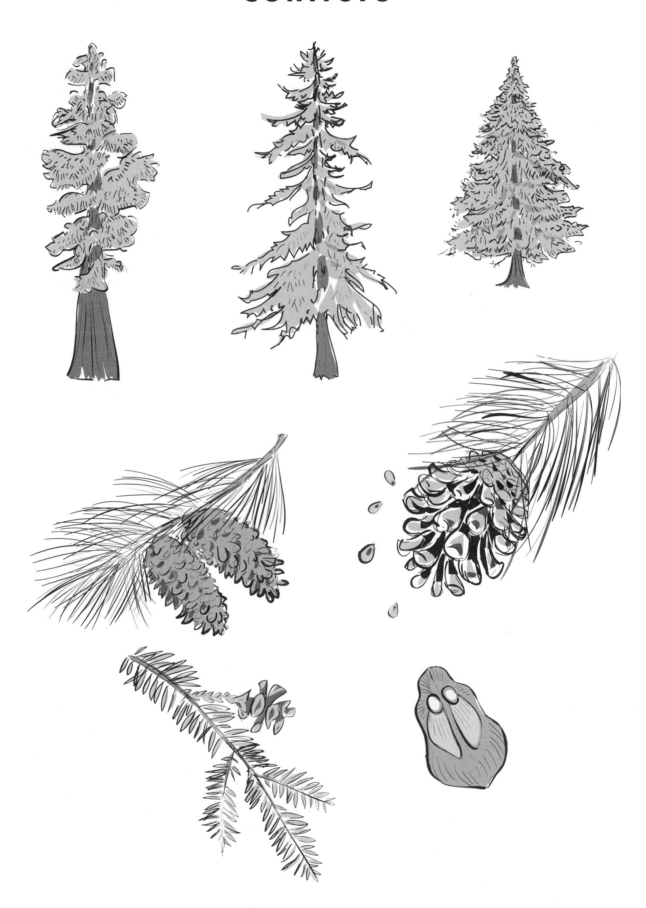

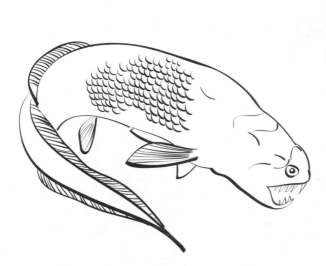

Armored Fish

Amphibians
(Salamander, Frog, Toad)

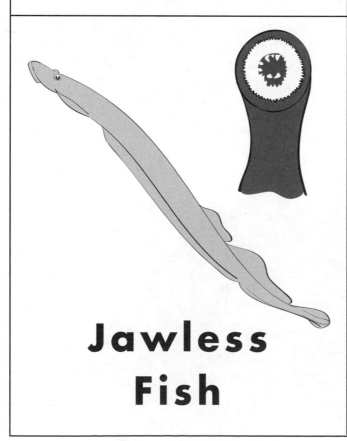

Jawless Fish

Sharks, Rays

Conifers

Major Evolutionary Events—Time Period #3

- **First amphibians appear on Earth.**

- **Fish are abundant in the seas.**

- **Amphibians dominate the land,** although they're awkward walkers. They develop eyelids for living out of water, but they still depend on water in order to take in oxygen through their skin, and because their eggs are jelly-like.

- **First forests form.**

- **Cone-bearing plants evolve,** with watertight seeds that retain moisture as the embryo forms. They also have thin, tough-coated needles that resist evaporation which allows them to invade drier parts of the land.

- **First reptiles begin to evolve** at the end of this period.

- **The supercontinent Pangaea forms** at the very end of this period.

Aquarium Background —Time Period #3 • *Session 4*

Fossils—Time Period #3

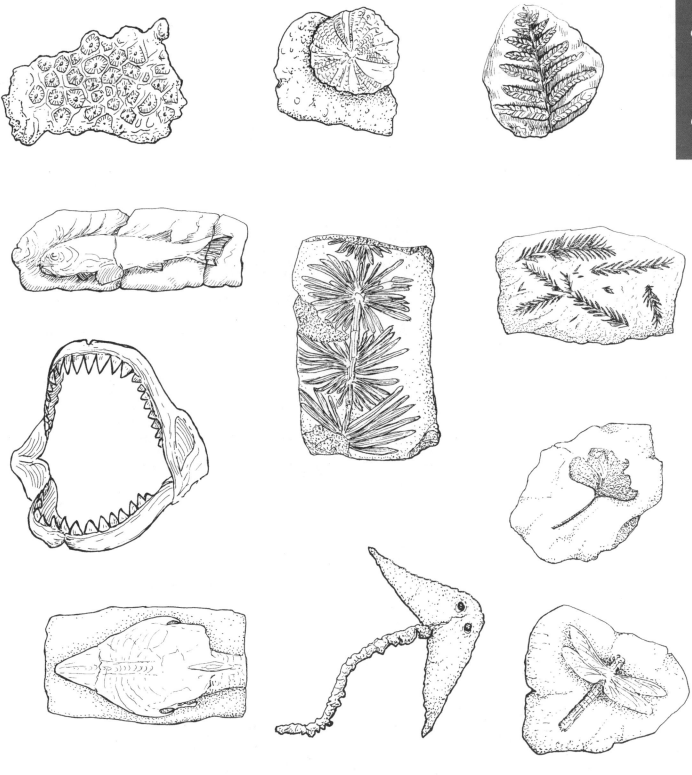

Continental Drift—Time Period #3

410–286 MYA

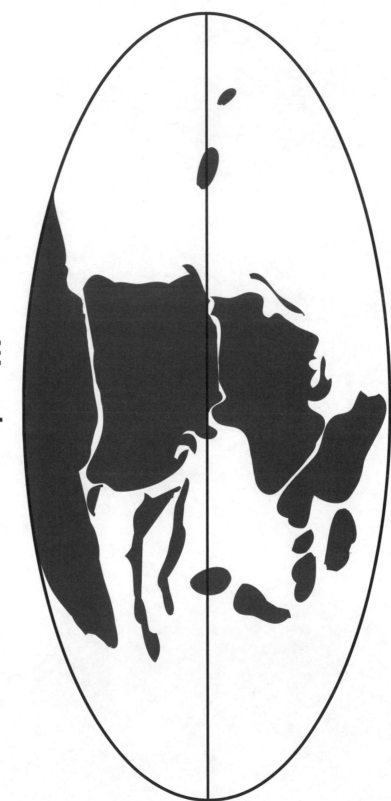

Weather

The weather will be pleasantly warm. The continent Laurasia is covered by warm, shallow seas. Excellent for diving! In some parts of the world the weather will be wet. No, the weather will be dry. Actually, the weather will be wet and dry, back and forth.

If you're planning to visit Australia, Africa, India, or South America, dress warmly. They are all at the South Pole, so they're mostly covered with ice.

Atmosphere

Mostly nitrogen and oxygen.

410 million to 286 million years ago

The Age of _____

Draw and label animals and plants you identify.

Time Travel Aquarium

Time Travel Terrarium

Fossils

Label any fossils you're able to identify.

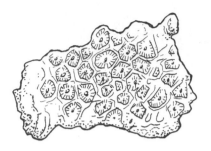

_____

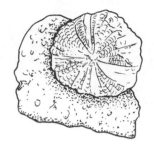

_____

_____

_____

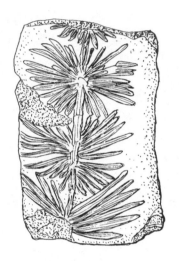

_____

_____

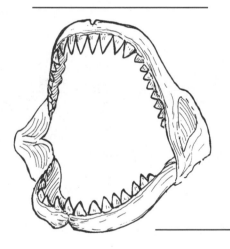

_____

_____

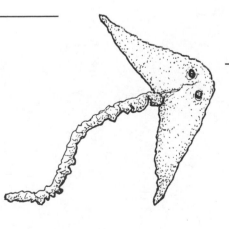

_____

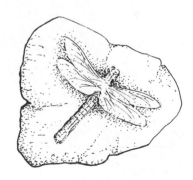

_____

© 2003 The Regents of the University of California. May be duplicated for classroom or workshop use.

Fish

Look at the pictures and write two or more statements about what these organisms have in common and how they are different.

_____

_____

_____

_____

_____

Amphibians

Look at the pictures and write two or more statements about what these organisms have in common and how they are different.

_____

_____

_____

_____

_____

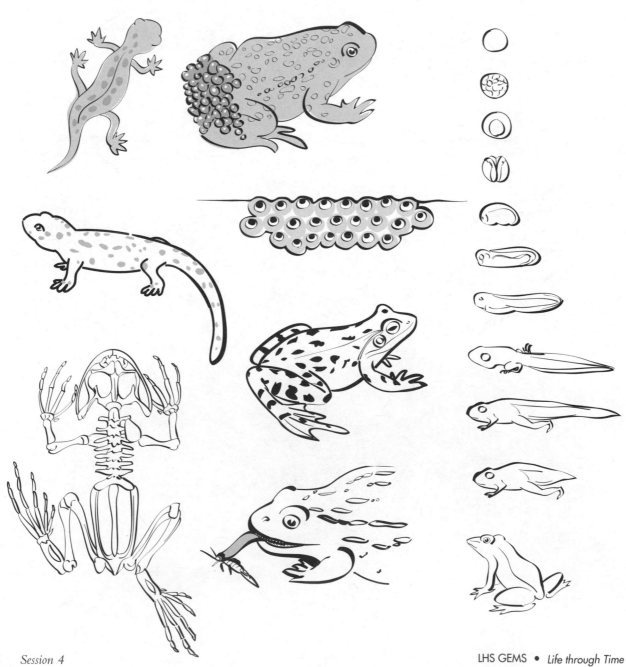

Conifers

Look at the pictures and write two or more statements about what these organisms have in common and how they are different.

_____

_____

_____

_____

_____

Organism Adaptations

Carefully observe the animals and plants, and write your best guesses to these questions:

Animal	What does it eat?	What eats it?	How does it move around?	How does it protect itself?

Plant	How does it feel?	What kind of environment does it live in? (Wet? Dry? Very dry?)	How does it reproduce?

Continental Drift

Draw the continents as they appeared in this time period, referring to the station sheet. Describe the weather and atmosphere.

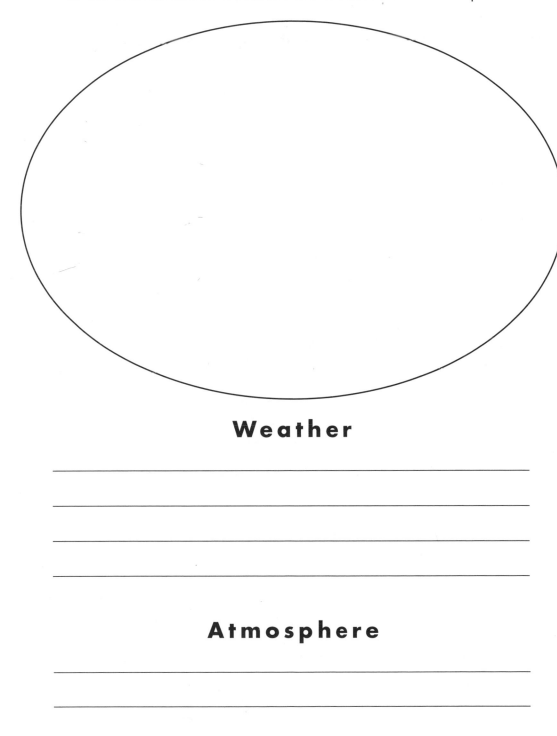

Weather

_____

_____

_____

_____

Atmosphere

_____

_____

_____

_____

Predictions for Next Time Period

Write your predictions for how organisms, land, water, weather, and atmosphere may change in the next time period.

_____

_____

_____

_____

_____

_____

_____

_____

_____

_____

_____

_____

_____

_____

_____

_____

Aquarium

1. *Archelon*
   (Giant Sea Turtle)
2. Ichthyosaurs
3. Crinoids
4. *Metriorhynchus*
   (Early Seagoing
   Crocodile)
5. Ammonites
6. Jelly

Terrarium

1. Ginkgo
2. *Rhamphorynchus*
   (Pterosaur)
3. Monkey Puzzle Trees
4. *Henkelotherium*
   (Early Mammal)
5. Butterfly
6. *Gallimimus bullatus*
7. Cycads
8. *Williamsonia*
9. *Barosaurus lentus*
10. Conifers
11. *Compsognathus*
12. *Archaeopteryx*
    (Early Bird)

Overview

The time is 286–65 MYA. As this evolution-rich period begins, the shallow seas that covered North America and many other parts of the world for the past 4,214 million years have been destroyed as the continents joined together to form the giant land mass Pangaea at the end of the last age. With the disappearance of these warm shallow seas, many of their resident organisms are now extinct. Movements of continents have blocked or diverted major ocean currents, resulting in the beginnings of global weather changes. Over the course of this time period, Pangaea will start to split apart into two new supercontinents. As dinosaurs and their kin (especially reptiles) diversify, they'll have a huge impact on the land, seas, and air.

By the end of the age, flowering plants will be flourishing, many animals—including the dinosaurs—will have gone extinct, and the first small and furry mammals will begin to exploit habitats vacated by the dinosaurs. Land will have separated into several distinct continents, pretty much as we know them today.

While what scientists call "background extinctions" occur every day due to habitat loss and over-harvesting of species, the history of life is also punctuated by several major disruptions known as "mass extinctions." Two of the most extraordinary of these occurred during the period covered by this session: the **Permian Extinction,** which wiped out 96 percent of all species in the ocean when warm, shallow seas disappeared, and the **KT Event,** which spelled the end of the dinosaurs and the rise and diversification of mammals. These are without a doubt the two most significant evolutionary events of this period.

In addition to the evolving core stations, this session features two evolutionary emblems of the period: reptiles and flowering plants. The wrap-up discussion focuses on reptiles and pollination. As before, students elect the most representative organism of the age; the **Life through Time** and **Tree of Life** wall charts and the **Class Time Line** are added to again; and students make predictions for what will occur in the next time period.

This session represents the Permian Period of the Paleozoic Era, and the Triassic, Jurassic, and Cretaceous Periods of the Mesozoic Era, all of the Phanerozoic ("Evident Life") Eon.

The term "Dinosauria" was coined by Sir Richard Owen in 1842 to describe the three "Fearfully Great Reptiles," or dinosaurs, known about at that time. Today, there's evidence that some dinosaurs may have been homeothermic ("warm-blooded"), like today's birds. See "Background for the Teacher" on page 270, for more on this.

The geologic record shows that a giant asteroid or meteorite slammed into the Earth 65 million years ago. A majority of scientists think that the impact of this collision shrouded the Earth in dust and debris for months, blocking sunlight and causing global temperatures to plummet. This, among other contributing factors, is thought to have caused the extinction of the dinosaurs. (Your students will have the opportunity to come up with this theory themselves, in Session 7.) The KT Event was named for the end of the Cretaceous (K, from the German Kreide) and Tertiary (T) periods.

■ What You Need

For the teacher:
- ❏ the appropriate **Fossils—Teacher's Answer Sheet** for this session (page 67), from your set

For the class:
- ❏ 1 copy of this session's **Most Representative Organism Script** (pages 186–187)
- ❏ 1 overhead transparency of **Flowering Plant Reproduction** (page 188)
- ❏ 1 overhead transparency of **Reptiles** (page 189)
- ❏ 1 overhead transparency of **Flowering Plants** (page 190)
- ❏ the overhead transparencies **Algae Reproduction, Moss Reproduction, Conifer Reproduction,** and **Amphibians** from previous sessions (pages 71, 119, 154, and 156)
- ❏ 1 copy of this session's **Tree of Life Organism Cards** (pages 191–192) to add to the **Tree of Life** wall chart
- ❏ 1 copy of **The Age of** _____ sign (page 78), if you're using it
- ❏ 1 copy of **Major Evolutionary Events—Time Period #4** (page 193)
- ❏ the adding-machine-tape strip you cut for the next session, "65 MYA–present time," to add to the **Class Time Line**
- ❏ 1 hard-shelled egg—a chicken egg will do (We recommend hard-boiled!)
- ❏ the **Time Travel Aquarium** and **Time Travel Terrarium** from the previous session
- ❏ the two large cloths from the previous session, to cover the aquarium and terrarium
- ❏ an overhead projector and screen

For the Time Travel Aquarium station:
- ❏ 1 copy of **Aquarium Background—Time Period #4** (page 200A)
- ❏ enough dechlorinated water to top off the aquarium, if needed
- ❏ plastic crocodiles, turtles, and plastic prehistoric marine reptiles—pleisosaurs, ichthyosaurs, mososaurs, etc.

For the Time Travel Terrarium station:
- ❏ 1 copy of **Terrarium Background—Time Period #4** (page 200B)

❑ any plastic dinosaurs and early terrestrial reptiles, including lizards, snakes, and tortoises

❑ one or two small flowering plants (such as sweet alyssum or any other easily available)

❑ beetles, ants, hissing cockroaches, or other real or plastic insects

❑ acorns

❑ plastic butterfly

❑ plastic or plush opossum

For the remaining core stations:

❑ 2 copies of the station sheet **Reptiles** (page 189)

❑ 2 copies of the station sheet **Flowering Plants** (page 190)

❑ 2 copies of the station sheet **Fossils—Time Period #4** (page 194)

❑ 3 copies of the station sheet **Continental Drift—Time Period #4;** two for the station and one for the **Life through Time** wall chart (page 195)

These are the six core stations for this session:
1) Time Travel Aquarium
2) Time Travel Terrarium
3) Continental Drift
4) Fossils
5) Reptiles
6) Flowering Plants

In addition to these six, you'll need at least four organism adaptations stations from the following list.

For the additional four or more stations:

❑ several containers to "house" organisms at the stations (Use whatever containers seem appropriate for the organism.)

❑ lizard

❑ snake

❑ turtle

❑ tortoise

❑ insects—especially those with complete, four-stage metamorphosis, such as ants, butterflies, and beetles

❑ flowering plants

❑ large flower(s) cuttings

❑ magnolia-tree flower or stem with leaves

❑ acorns
❑ cypress tree branch
❑ hickory nut
❑ reptile egg (do not collect viable eggs from the wild)
❑ shed reptile skin
❑ reptile skull or skeleton

For each student:
❑ **Time Travel Journal** from the previous session
❑ **Organism Key** from the previous session
❑ 1 set of **Time Travel Journal** pages labeled **Time Period #4** (pages 196–202; 7 pages total) to add to journal *(Don't copy these until you've added this session's organisms to the **Organism Adaptations** page; see "Getting Ready," below.)*

■ Getting Ready

Before the Day of the Activity

1. Copy and cut up this session's **Most Representative Organism Script** (pages 186–187), and contact adult volunteers.

2. Write the names of the organisms you've acquired for this session in the left-hand column on the **Organism Adaptations** page of the students' journals.

3. Copy the **Time Travel Journal** pages marked **Time Period #4** (pages 196–202; 7 pages total), punching holes if needed.

4. Copy and set aside the overhead transparencies for this session: **Flowering Plant Reproduction, Reptiles,** and **Flowering Plants** (pages 188–190).

5. Label the containers you'll be using for the organisms at the **Organism Adaptations** stations. Label them "Reptiles" or "Flowering Plants," then more specifically by kind, such as "lizard" or "magnolia."

6. Copy and cut up this session's **Tree of Life Organism Cards** (pages 191–192), ready to place on the **Tree of Life** wall chart.

7. Update the **Time Travel Aquarium.**

   a. Top off the water level if needed.

b. Lightly tape **Aquarium Background—Time Period #4** (page 200A) to the back of the aquarium.

c. Add the plastic crocodile and any other plastic marine reptiles you've acquired.

8. Update the **Time Travel Terrarium.**

a. Lightly tape **Terrarium Background—Time Period #4** (page 200B) to the back of the terrarium.

b. Plant any flowering plants or cuttings in the soil.

c. Add any plastic dinosaurs and early terrestrial reptiles.

d. Add any real or plastic beetles, ants, or hissing cockroaches; plastic butterflies; plastic or plush opossum; and acorns you may have acquired.

9. Copy and set aside the **Major Evolutionary Events—Time Period #4** sheet (page 193), ready to post over the **Class Time Line.**

On the Day of the Activity

1. Make copies of the station sheets and set up the remaining core stations.

a. Set out the two copies of **Reptiles** (page 189) at a station.

b. Set out the two copies of **Flowering Plants** (page 190) at a station.

c. Set out the two copies of **Fossils—Time Period #4** (page 194) at a station.

d. Set out two copies of **Continental Drift—Time Period #4** (page 195) at a station.

2. Place the animals and plants in their containers and set up the additional stations.

3. Gather the students' **Time Travel Journals** and **Organism Keys** from the previous session, ready to distribute. Have the appropriate **Fossils—Teacher's Answer Sheet** for this session (page 67) readily available.

4. Put out sets of **journal pages** labeled **Time Period #4** for students to add to their journals.

5. Set aside the third copy of the **Continental Drift—Time Period #4** sheet to add to the **Life through Time** wall chart.

6. Copy **The Age of _____** sign (page 78) or write the information on a blank 8 ½" x 11" sheet of paper.

7. Set out the overhead transparencies **Algae Reproduction, Moss Reproduction, Conifer Reproduction,** and **Amphibians** from previous sessions (pages 71, 119, 154, and 156).

8. Cover the aquarium and terrarium.

9. Set aside the strip "65 MYA–present time," ready to add to the **Class Time Line.**

■ Prepare for and Begin the Time Travel

1. Announce to your students the new time period: 286–65 million (286,000,000–65,000,000) years ago, or 286–65 MYA. Write it on the board or point it out on the previous session's **Class Time Line.** This period lasted 221 million (221,000,000) years.

2. Distribute the **Time Travel Journals,** the sets of new journal pages for this time period, and the **Organism Keys.** Ask students to review their predictions for what the Earth will be like during this time period, including land, water, plants, animals, weather, atmosphere, and continents.

3. Briefly discuss some of the students' predictions. Accept all answers.

4. If needed, briefly review the standard stations, and the two new ones for this session—reptiles and flowering plants.

5. Divide students into teams of two and begin the rotations.

**Stations recap—
Session 5/Time Period #4**

1) Time Travel Aquarium
2) Time Travel Terrarium
3) Continental Drift
4) Fossils
5) Reptiles
6) Flowering Plants
7) through 10)—whatever additional stations you've chosen

■ Time Travel Debrief

1. Focus the students on the **Life through Time** wall chart. Be sure they have their **Time Travel Journals** and **Organism Keys** for the discussion.

2. Remove the background illustrations from the aquarium and terrarium and place them on the **Life through Time** wall chart.

3. Debrief the **Aquarium, Terrarium,** and **Fossils** stations as before.

4. Debrief the **Reptiles** station.

Put the **Reptiles** transparency on the projector. Ask a few students to share their statements about reptiles. Be sure the following important points are brought out:

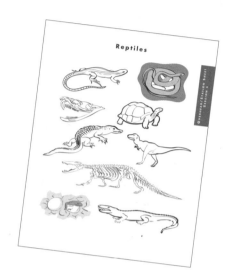

Reptiles

- **Reptiles are ectothermic.** (Review this, if needed. Ectothermic animals' internal temperature depends on the temperature outside; they can't maintain a constant internal temperature. This means they can't generate their own heat when the surrounding temperature is low, or cool down when it's hot.)

- **Reptiles have watertight skin** (which retains moisture) **and scales** (which protect from the environment). These adaptations allow reptiles to live in dry areas—but the hardness of their skin and scales is also confining; reptiles, like animals with exoskeletons, need to molt in order to grow.

- **They lay leathery, watertight eggs,** which also allows them to reproduce in dry areas. (Show the class the hard-shelled egg, and explain that even though it's a bird egg, watertight eggs first developed in reptiles. Ask your students if this gives them an idea from what animals birds may have evolved.)

- **They have backbones** made of bone, not cartilage.

Do birds come from dinosaurs?

Many paleontologists, taxonomists, and other scientists who study this question continue to feel that there is compelling evidence that birds evolved from small, terrestrial theropod ("beast-footed") dinosaurs, in the group (clade) Maniraptora. See "Background for the Teacher" on page 270 for more on this.

- **They have teeth;** not for chewing, but for grabbing food.

- **Reptiles were more complex than earlier life,** in keeping with the evolutionary pattern.

5. Ask a few students to share their statements about what the organisms from the **Amphibians** sheet had in common with the organisms from the **Reptiles** sheet. Alternate the **Amphibians** and **Reptiles** transparencies on the overhead projector during the discussion. Make sure these points are made:

Early Amphibian

Reptile

What Reptiles and Amphibians Have in Common

- **A backbone.** Their skeletons match up, bone type for bone type.

- **Teeth;** not for chewing, but for holding on to food.

- **A large intestine** for extracting water from waste products.

- **"One-way traffic" guts.**

6. Debrief the **Organism Adaptations** stations.

Have students refer to the Organism Adaptations pages in their journal. Revisit each animal and plant from the stations, asking the students for a few of their ideas on each organism's adaptations.

7. Debrief the **Flowering Plants** station.

a. Show the overhead transparency **Flowering Plants** and ask a few students to share what the flowering plants on the sheet had in common.

b. Review the reproductive strategies of algae, moss, and conifers, using the overhead transparencies **Algae Reproduction, Moss Reproduction,** and **Conifer Reproduction** from previous sessions. Ask your students how they think the newly evolved flowering plants reproduced, compared to the algae and earlier plants.

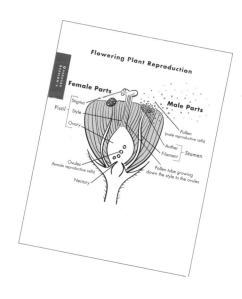

c. Show this session's **Flowering Plant Reproduction** transparency. Point out the male and female parts of the flower. Explain that, originally, flowers were probably all wind-pollinated, like cone-bearing plants. (Some flowering plants, like grasses, are still wind-pollinated today.)

d. Ask students what they think some of the problems with wind pollination might be. [Wind is not consistent; much pollen is likely to be blown into areas where there are no other plants to pollinate; throwing pollen out to the wind is wasteful—insects probably eat a lot of it, but that doesn't do the plants any good!]

e. Ask students how they think many flowering plants evolved away from wind pollination. [When insects and other organisms brushed up against or fed from the flowers, the pollen "stuck" to them and traveled with them from plant to plant. With the success of this accidental pollination system, flowers began to evolve in ways that made this kind of pollination more likely.]

f. Ask students in what ways flowers may have evolved to attract these insect pollinators. Explain that flowers with bright colors and tasty nectar turned out to be the most successful "attractors." Over millions of years, the flowers and pollinators became more and more specialized for each other. In many flowering plants, such as orchids, the flower head even evolved a more "pollinator friendly" (or pollinator trapping!) structure.

There are many excellent books that describe the amazing architectural "tricks" that flowers play on their pollinators to ensure reproduction. See "Resources" on page 307 for some suggestions.

8. Debrief the **Continental Drift** station.

a. Review how the continents appeared during the previous time period on the **Life through Time** wall chart and ask students how the continents moved during the time period they just studied. Be sure they raise the following points:

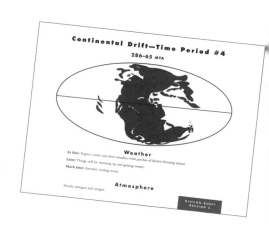

- At the beginning of the period, Pangaea is still intact and drifting north, so glaciers are retreating southward.
- Some continents begin to separate from Pangaea and drift apart.

- By the end of this period, Pangaea has split into two new supercontinents—again named Gondwana (now in the Southern Hemisphere) and Laurasia (to the north), but made up of different land masses than before.
- The temperature has so warmed up that there are no longer any ice caps on the planet, even at the poles.

b. Add the station sheet **Continental Drift—Time Period #4** to the **Life through Time** wall chart.

■ Major Evolutionary Events

1. Ask students to consider all they've just discussed, then brainstorm some major evolutionary changes that took place in the organisms or their habitats during the period they just studied, referring to the list below as a content checklist.

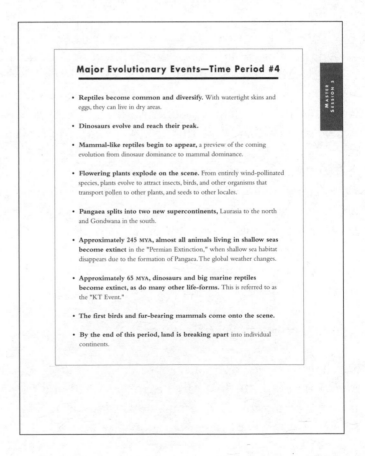

Major Evolutionary Events—Time Period #4

- **Reptiles become common and diversify.** With watertight skins and eggs, they can live in dry areas.

- **Dinosaurs evolve and reach their peak.**

- **Mammal-like reptiles begin to appear,** a preview of the coming evolution from dinosaur dominance to mammal dominance.

- **Flowering plants explode on the scene.** From entirely wind-pollinated species, plants evolve to attract insects, birds, and other organisms that transport pollen to other plants, and seeds to other locales.

- **Pangaea splits into two new supercontinents,** Laurasia to the north and Gondwana in the south.

- **Approximately 245 MYA, almost all animals living in shallow seas become extinct** in the "Permian Extinction," when shallow sea habitat disappears due to the formation of Pangaea. The global weather changes.

- **Approximately 65 MYA, dinosaurs and big marine reptiles become extinct, as do many other life-forms.** This is referred to as the "KT Event."

- **The first birds and fur-bearing mammals come onto the scene.**

- **By the end of this period, land is breaking apart** into individual continents.

2. Ask students how their observations compare with the predictions they made in Session 4.

3. Post the sheet **Major Evolutionary Events—Time Period #4** above this session's strip of the **Class Time Line.**

■ Most Representative Organism of the Age

1. Tell students it's time to name this time period, and remind them they'll get to vote after each candidate has made a speech.

2. Hand out the cut-up speeches to those who'll play each role, and begin.

3. After the speeches, ask if anyone wishes to nominate and make an improvised speech for any other organism.

4. Reconvene the class and review the candidates with the class:

    • The Age of *Tyrannosaurus rex* (or the Age of Dinosaurs)
    • The Age of Reptiles
    • The Age of Flowering Plants

5. Conduct and tally the vote. Write the name of the winner on the **Age of**_____ sign, a blank sheet, or directly on the wall chart. Have students write it in their journals. If you made a sign, add it to the chart.

Note: The chart should now be arranged as shown here.

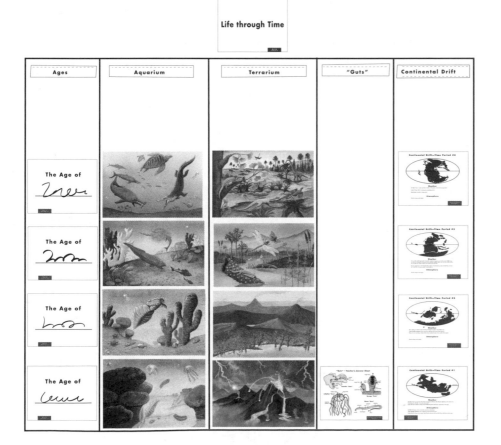

■ Class Time Line

1. Attach to the **Class Time Line** the strip of adding-machine tape you pre-cut for the next session, "65 MYA–present time," and let your students know this is the period they'll focus on next.

2. Write the years of the next time period on the board:

   - 65 million years ago–present time
   - 65,000,000 years ago–present time
   - 65 MYA–present time

3. Have students write in their **Time Travel Journals** their predictions about the organisms and habitat changes they think they'll encounter in the next time period.

4. Ask students to bring in organisms you'll need for the next session's organism adaptation stations.

5. If there's time, have one or more students place this session's **Tree of Life Organism Cards** on the **Tree of Life** wall chart as shown, or do this yourself before the next session.

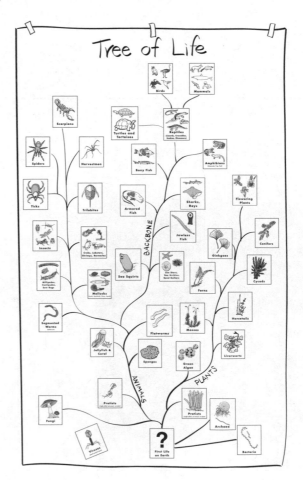

■ Going Further

1. Incorporating Related Activities

Any activities you have from other curricula relating to pollination, reptiles, dinosaurs, or eggs would be an excellent addition at this point.

2. Songs

We Don't Wanna Go (The Dinosaur Graduation Song), on page 327, deals with dinosaur extinction. *One Of These Days We're Gonna Rule The World (The Cockroach),* a hard rock tune (page 328), and the *Family Tree* song (page 329), both provide overall review of general evolutionary trends. You can preview these songs at www.moo-boing.com.

Script for Most Representative Organism Election for
Time Period #4: 286–65 MYA

1

Tyrannosaurus rex: *[Loud and bossy.]* RRROOOOOAAAAAARRRR! No question about it, this was the "Age of the Dinosaurs," and more specifically, the age of **TYRANNOSAURUS REX** (but you can call me *T. rex*). As my name tells you, I am the *king* of the tyrant lizards. Dinosaurs took over the world during this time, and we ruled for 150 million years. At 48 feet long, and 20 feet tall, I was the king of them all! *[Said sweetly.]* So **please** be so kind as to vote for me. I would be ever so grateful. *[Yelling again.]* Ahhh to heck with it, I'm *T. rex*—vote for me or else! ARRRRGHHHH!

2

Lizard: *[To T. rex.]* You don't scare me—you're extinct! Yep ladies and gentlemen, *T. rex* has left the building. And this should *not* be called the "Age of Dinosaurs." (By the way, *T. rex* wasn't even around for most of this age, so was definitely not "the king.") Don't get me wrong, dinosaurs were certainly the biggest animals back then, but they were only **one kind** of reptile. Hey, some of my best friends are dinosaurs *[Looks at* T. rex.*]*...or I guess **were** dinosaurs *[Laughs at* T. rex.*]*. But other reptiles like turtles, crocodiles, lizards, and snakes evolved at this time. And **we** didn't go extinct. It should be called the "Age of Reptiles."

3

Magnolia: OK, I'm not loud like a *T. rex,* and not fast like a lizard. I'm a flower, and this should be called the "Age of Flowering Plants." Why? Because all life depends on plants, and us flowering plants took over during this time. The key to our success, you ask? **Pollination.** You see, when it's time to mate, plants can't walk over to each other like animals. We started out with wind pollination, letting the wind carry our pollen from one flower to another. Then we discovered how easy it is to sucker insects, birds, and bats in to do the work for us. Just a little nectar, a little bright color to advertise, and here they come. Vote for us!

Flowering Plant Reproduction

Female Parts

Male Parts

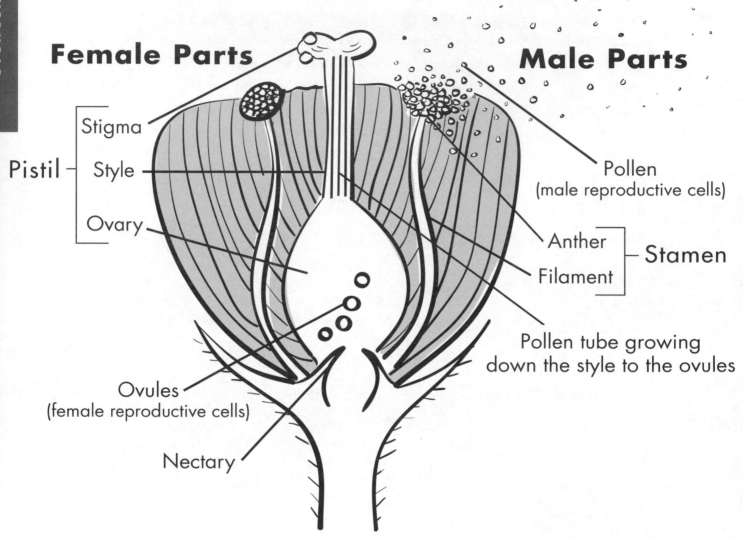

Pistil {
- Stigma
- Style
- Ovary

Ovules
(female reproductive cells)

Nectary

Pollen
(male reproductive cells)

Anther
Filament
} Stamen

Pollen tube growing
down the style to the ovules

Reptiles

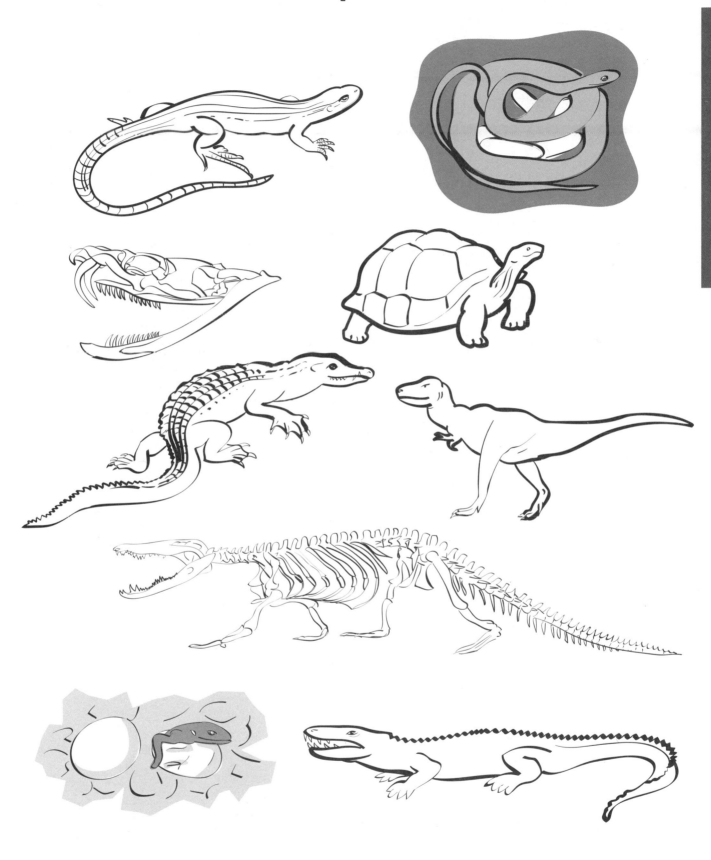

Flowering Plants

Iris

Poppy

Grass

Orchid

Cactus

Magnolia

Dogwood

Oak

Apple

Tomato

Pumpkin

Sunflower

Bony Fish

Turtles and Tortoises

Flowering Plants

Reptiles
(Lizards, Crocodiles, Snakes, Dinosaurs)

Tree of Life Organism Cards
Time Period #4

Birds

Mammals

Major Evolutionary Events—Time Period #4

- **Reptiles become common and diversify.** With watertight skins and eggs, they can live in dry areas.

- **Dinosaurs evolve and reach their peak.**

- **Mammal-like reptiles begin to appear,** a preview of the coming evolution from dinosaur dominance to mammal dominance.

- **Flowering plants explode on the scene.** From entirely wind-pollinated species, plants evolve to attract insects, birds, and other organisms that transport pollen to other plants, and seeds to other locales.

- **Pangaea splits into two new supercontinents,** Laurasia to the north and Gondwana in the south.

- **Approximately 245 MYA, almost all animals living in shallow seas become extinct** in the "Permian Extinction," when shallow sea habitat disappears due to the formation of Pangaea. The global weather changes.

- **Approximately 65 MYA, dinosaurs and big marine reptiles become extinct, as do many other life-forms.** This is referred to as the "KT Event."

- **The first birds and fur-bearing mammals come onto the scene.**

- **By the end of this period, land is breaking apart** into individual continents.

Fossils—Time Period #4

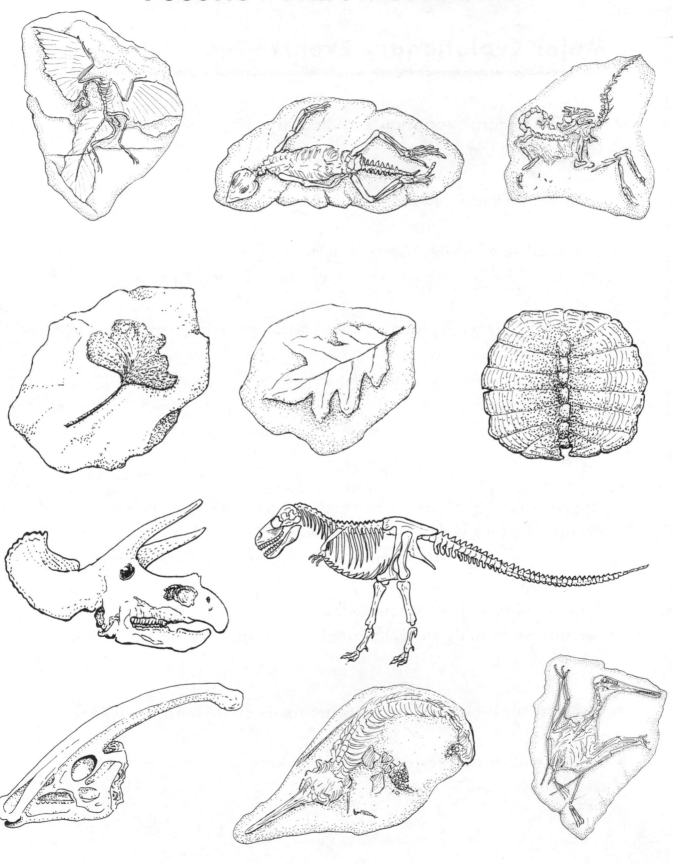

Continental Drift—Time Period #4

286–65 MYA

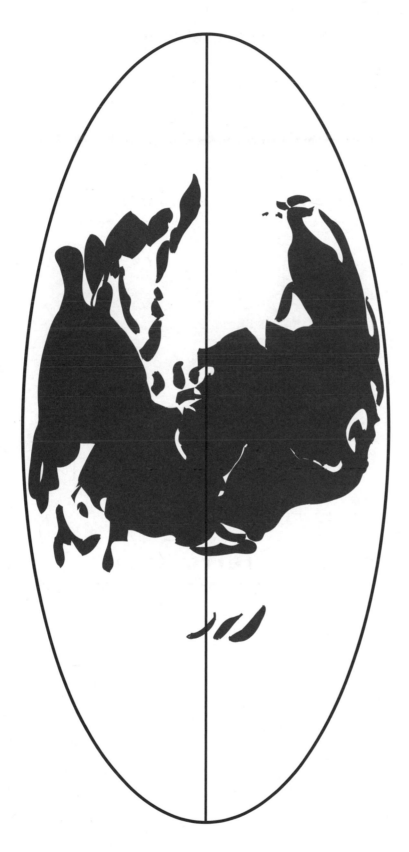

Weather

At first: Expect cooler and drier weather, with patches of deserts forming inland.

Later: Things will be warming up and getting wetter.

Much later: Another cooling trend.

Atmosphere

Mostly nitrogen and oxygen.

Time Travel Journal —Time Period #4

286 million to 65 million years ago

The Age of _____

Draw and label animals and plants you identify.

Time Travel Aquarium

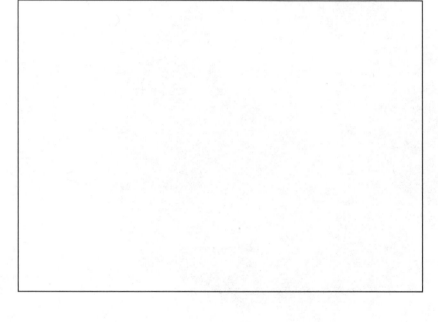

Time Travel Terrarium

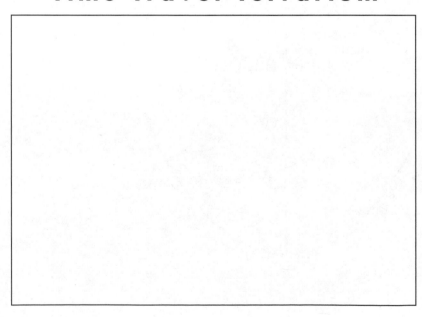

Fossils

Label any fossils you're able to identify.

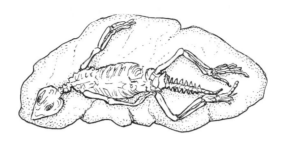

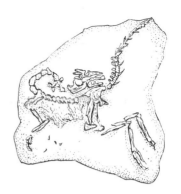

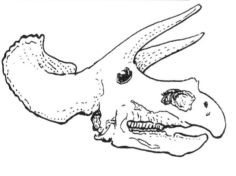

_____   _____   _____

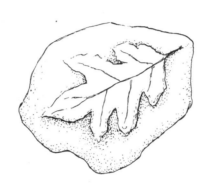

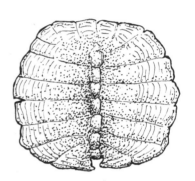

_____   _____   _____

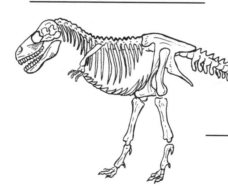

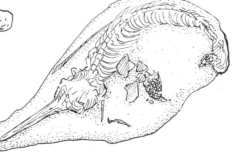

_____   _____

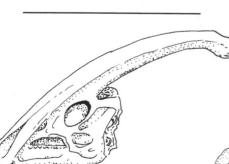

_____   _____   _____

Reptiles

Look at the pictures and write two or more statements about what these organisms have in common and how they are different.

_____

_____

_____

_____

_____

Flowering Plants

Look at the pictures and write two or more statements about what these organisms have in common and how they are different.

_____

_____

_____

_____

_____

Time Travel Journal —Time Period #4

Organism Adaptations

Carefully observe the animals and plants, and write your best guesses to these questions:

Animal	What does it eat?	What eats it?	How does it move around?	How does it protect itself?

Plant	How does it feel?	What kind of environment does it live in? (Wet? Dry? Very dry?)	How does it reproduce?

Aquarium Background —Time Period #4   •   *Session 5*

Terrarium Background —Time Period #4   •   *Session 5*

Continental Drift

Draw the continents as they appeared in this time period, referring to the station sheet. Describe the weather and atmosphere.

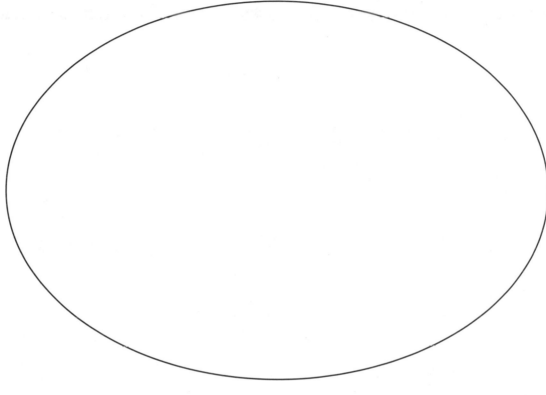

Weather

_____

_____

_____

_____

Atmosphere

_____

_____

_____

_____

Predictions for Next Time Period

Write your predictions for how organisms, land, water, weather, and atmosphere may change in the next time period.

_____

_____

_____

_____

_____

_____

_____

_____

_____

_____

_____

_____

_____

_____

_____

_____

Aquarium

1. Puffins
2. *Hesperornis*
   (Flightless Diving Birds)
3. Bony Fish
4. *Basilosaurus*
   (Early Whale)
5. Mackerel
6. Jelly

Terrarium

1. Conifers
2. Condors
3. Ducks
4. *Australopithecus afarensis*
   (Early Hominid)
5. *Smilodon*
   (Saber-Toothed Cat)
6. Grassy Meadow
7. Deciduous Trees
8. Magnolias
9. *Diatryma*
   (Flightless Predatory Bird)
10. *Hyracotherium*
    (Early Horses)
11. Platybelodon
    ("Shovel-Tusk" Elephant)
12. Cattails
13. Water Lilies

Overview

We're creeping up on the present! During the final time period to be explored, the students visit the planet as it evolved from 65 million years ago to this very day; 65 MYA–present time.

With the evolution of flowering plants and the disappearance of the dinosaurs and many predatory reptiles, the warm-blooded, furry mammals diversify. Grasses, a new type of flowering plant, appear, leading to the evolution of many grazing mammals. In the most recent and teensiest span of this period, humans at last appear on the scene. The continents are drifting into their present positions, and the climate is cooling.

This session's time-travel stations, the last of the unit, include the Time Travel Aquarium and Terrarium, organisms representing the time period, Continental Drift, Fossils, "Guts" revisited, and two new ones: Birds, and Meat- and Plant-Eating Mammals. Students elect the most representative organism of the time, and the **Life through Time** wall chart and **Class Time Line** are brought up to the present.

> We're now in the Cenozoic Era, the era of "Recent Life." The Tertiary and Quaternary Periods and a good many epochs make up this era; see page 270 of "Background for the Teacher."

■ What You Need

For the teacher:
- ❏ the appropriate **Fossils—Teacher's Answer Sheet** for this session (page 68), from your set

For the class:
- ❏ 1 copy of this session's **Most Representative Organism Script** (page 218)
- ❏ 1 overhead transparency of **Birds** (page 219)
- ❏ 1 overhead transparency of **Mammals** (page 220)
- ❏ 1 overhead transparency of **Skeletons through Time** (page 221)
- ❏ 1 overhead transparency of **"Guts": Bird and Mammal** (page 222)
- ❏ the overhead transparencies **Algae Reproduction, Moss Reproduction, Conifer Reproduction, Flowering Plant Reproduction, Reptiles,** and **"Guts": Single-Celled Organism/Sponge/Jellyfish/Earthworm** from previous sessions (pages 71, 119, 154, 188, 189, and 72)

❑ 1 copy of **The Age of** _____ sign (page 78), if you're using it
❑ 1 copy of **Major Evolutionary Events—Time Period #5** (page 223)
❑ the **Time Travel Aquarium** and **Time Travel Terrarium** from the previous session
❑ the two large cloths from the previous session, to cover the aquarium and terrarium
❑ an overhead projector and screen

For the Time Travel Aquarium station:
❑ 1 copy of **Aquarium Background—Time Period #5** (page 224A)
❑ enough dechlorinated water to top off the aquarium, if needed
❑ plastic whales and other plastic marine mammals
❑ plastic sharks, including a large plastic great white shark *(Carcharodon carcharias)* model if you have one (to represent the giant *Carcharodon megalodon,* related to the present-day great white)

For the Time Travel Terrarium station:
❑ 1 copy of **Terrarium Background—Time Period #5** (page 224B)
❑ any plastic early mammals (wooly mammoth, saber-toothed cat, etc.)
❑ any plastic modern mammals, birds, and other animals (but not domesticated animals such as dogs)
❑ more small flowering plants
❑ grasses

For the remaining core stations:
❑ 2 copies of the station sheet **Birds** (page 219)
❑ 2 copies of the station sheet **Mammals** (page 220)
❑ 3 copies of the station sheet **"Guts": Bird and Mammal** (page 222)
❑ 2 copies of the station sheet **Fossils—Time Period #5** (page 224)
❑ 3 copies of the station sheet **Continental Drift—Time Period #5;** two for the station and one for the **Life through Time** wall chart (page 225)

These are the seven core stations for this session:
1) Time Travel Aquarium
2) Time Travel Terrarium
3) Continental Drift
4) Fossils
5) Birds
6) Mammals
7) "Guts"

In addition to these seven, you'll need at least three organism adaptations stations from the following list.

For the additional three or more stations:
- ❑ several containers to "house" organisms at the stations (Use whatever containers seem appropriate for the organism.)
- ❑ any safe mammal
- ❑ any safe bird
- ❑ bird skull
- ❑ mammal skull
- ❑ grasses

For each student:
- ❑ **Time Travel Journal** from the previous session
- ❑ **Organism Key** from the previous session
- ❑ 1 set of **Time Travel Journal** pages labeled **Time Period #5** (pages 226–232; 7 pages total) to add to journal *(Don't copy these until you've added this session's organisms to the **Organism Adaptations** page; see "Getting Ready," below.)*

■ Getting Ready

Before the Day of the Activity

1. Copy and cut up this session's **Most Representative Organism Script** (page 218), and contact adult volunteers.

2. Write the names of the organisms you've acquired for this session in the left-hand column on the **Organism Adaptations** page of the students' journals.

3. Copy the **Time Travel Journal** pages marked **Time Period #5** (pages 226–232; 7 pages total), punching holes if needed.

4. Copy and set aside the overhead transparencies for this session: **Birds, Mammals, Skeletons through Time,** and **"Guts": Bird and Mammal** (pages 219–222).

5. Label the containers you'll be using for the organisms at the **Organism Adaptations** stations. (First "Birds," "Mammals," or "Plants," then more specifically by kind, such as "parakeet," "hamster," or "alyssum.")

6. Update the **Time Travel Aquarium.**

   a. Top off the water level if needed.

   b. Lightly tape **Aquarium Background—Time Period #5** (page 224A) to the back of the aquarium.

   c. Remove any aquatic dinosaurs from Session 5, and add the plastic whales and other plastic marine mammals you've acquired.

   d. Add the plastic shark models.

7. Update the **Time Travel Terrarium.**

   a. Lightly tape **Terrarium Background—Time Period #5** (page 224B) to the back of the terrarium.

   b. Plant any additional flowering plants or cuttings you have, including grasses.

   c. Remove any terrestrial dinosaurs and early reptiles from the previous session and replace with any plastic early mammals, birds, and other modern animals.

8. Copy and set aside the **Major Evolutionary Events—Time Period #5** sheet (page 223), ready to post over the **Class Time Line.**

Note: We suggest you read ahead through Session 7 and choose which option you'll use to conclude the unit. At the end of Session 6, you can then ask students to bring in items and containers for the next session.

On the Day of the Activity

1. Make copies of the station sheets and set up the remaining core stations.

   a. Set out the two copies of **Birds** (page 219) at a station.

   b. Set out the two copies of **Mammals** (page 220) at a station.

   c. Set out two copies of **"Guts": Bird and Mammal** (page 222) at a station.

   d. Set out the two copies of **Fossils—Time Period #5** (page 224) at a station.

   e. Set out two copies of **Continental Drift—Time Period #5** (page 225) at a station.

2. Place the animals and plants in their containers and set up the additional stations.

3. Gather the students' **Time Travel Journals** and **Organism Keys** from the previous session, ready to distribute. Have the appropriate **Fossils—Teacher's Answer Sheet** for this session (page 68) readily available.

4. Put out sets of **journal pages** labeled **Time Period # 5** for students to add to their journals.

5. Set aside the third copies of the **"Guts": Bird and Mammal** and **Continental Drift—Time Period #5** sheets to add to the **Life through Time** wall chart.

6. Copy **The Age of** _____ sign (page 78) or write the information on a blank 8 ½" x 11" sheet of paper.

7. Set out the overhead transparencies **Algae Reproduction, Moss Reproduction, Conifer Reproduction, Flowering Plant Reproduction, Reptiles,** and **"Guts": Single-Celled Organism/ Sponge/Jellyfish/Earthworm** from previous sessions (pages 71, 119, 154, 188, 189, and 72).

8. Cover the aquarium and terrarium.

■ **Prepare for and Begin the Time Travel**

1. Tell your students that this is the last time period they'll be traveling to. It runs from 65 million (65,000,000) years ago to the present time, or 65 MYA to this very day. Write the time period on the board or point it out on the previous session's **Class Time Line.**

2. Ask the students if they can guess what significant species will make its first appearance on Earth in the very, very last part of this session (the very, very recent past). [Humans! We've been on Earth for just the blink of an eye, in geologic terms. This is an important concept for students to learn.]

3. Distribute the **Time Travel Journals,** the sets of new journal pages for this time period, and the **Organism Keys.** Ask students to review their predictions for what the Earth will be like during this time period, including land, water, plants, animals, weather, atmosphere, and continents.

4. Briefly discuss some of the students' predictions. Accept all answers.

5. If needed, briefly review the standard stations, and the three new ones for this session—birds, mammals, and "guts."

6. Divide students into teams of two, assign each team to its first station, and let them rotate through the others at their own pace and in any order.

■ **Time Travel Debrief**

1. Focus the students on the **Life through Time** wall chart. Be sure they have their **Time Travel Journals** and **Organism Keys** for the discussion.

2. Remove the background illustrations from the aquarium and terrarium and place them on the **Life through Time** wall chart.

3. Debrief the **Aquarium, Terrarium,** and **Fossils** stations as before.

4. Debrief the **Birds** station.

**Stations recap—
Session 6/Time Period #5**

1) Time Travel Aquarium
2) Time Travel Terrarium
3) Continental Drift
4) Fossils
5) Birds
6) Mammals
7) "Guts"
8) through 10)—whatever additional stations you've chosen

Put the overhead transparency **Birds** on the projector. Ask a few students to share their statements about birds. Be sure these points are made:

Birds

- **They're endothermic,** which allows them to be active in cold areas. (Review this, if needed. Endothermic animals can maintain a constant internal temperature. This means they can generate their own heat when the surrounding temperature is low, and cool down when the temperature is high.)

- **They have feathers,** which greatly enhance flight and provide excellent insulation for roosting in cold areas or at night; swimming in cold water (penguins and murres, for instance); or flying at altitudes where the temperature is low.

- **They have a watertight, hard-shelled egg.** Birds have a common ancestor with reptiles, and have similar waterproof, shelled eggs. Birds evolved a hardening compound in their eggshell (phosphatized calcium carbonate), which increases the egg's strength.

- **Birds' skin is watertight.** This, and their watertight eggs, allowed them to retain moisture and invade drier parts of the land.

- **They have a backbone.**

- **Their bones are light and hollow,** which minimizes weight during flight, and **their bodies are rigid and fused** (like the rigid body of a plane), to help minimize stress on the skeleton during flight.

- **They have beaks in place of teeth.** (Beaks are lighter than teeth.) Since beaks aren't good for chewing, birds (like earthworms) have a gizzard, where food is crushed and ground up. Some swallow small rocks to help with the grinding. (That's what chickens and doves are doing when they peck in the gravel!)

- **Birds were more complex than earlier life-forms,** and recent birds are more complex than early birds.

---

5. Ask a few students to share their statements about what the organisms from the **Reptiles** sheet had in common with the organisms from the **Birds** sheet. Alternate the **Reptiles** and **Birds** transparencies on the overhead projector during the discussion. Make sure the students bring up or consider these points:

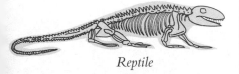
Reptile

Bird

What Reptiles and Birds Have in Common

- **A backbone.** Their skeletons match up, bone type for bone type.

- **A watertight, shelled egg.**

- **Watertight skin.**

- **A large intestine** for extracting water from waste products.

- **"One-way traffic" guts.**

- **Scaly skin (on legs in birds).**

- **They both molt (shed).** Birds molt feathers; reptiles molt their skins.

- **Feathers.** According to recent finds, some dinosaurs (reptiles) had feathers too.

6. Debrief the **Mammals** station.

Put the overhead transparency **Mammals** on the projector, and ask a few students to share their statements about mammals. Be sure the following points are made:

Mammals

- **They're endothermic,** which allows them to be active in cold areas.

- **They have hair** (even whales have some!), which serves as excellent insulation for cold areas.

- **They have mammary glands,** to feed their often defenseless young. (This adaptation allowed mothers to forage for food for themselves and then feed their young with the milk they produced.)

- **Most have teeth,** specialized to do different jobs:

  —In *meat-eaters,* most teeth are pointed, for tearing flesh.
  —In *plant-eaters,* some teeth are shaped for snipping plants, some for grinding.
  (In some whales, a fringe-like filter called **baleen** takes the place of teeth.)

- **Almost all mammals give birth to live young,** rather than laying eggs.

- **They have a backbone made of bone, not carti-lage.**

- **Mammals are more complex** than earlier life-forms, and recent mammals are more complex than early mammals.

---

*Mammals that bear live young are called **viviparous**. Egg-laying mammals are called **oviparous**. There are only two oviparous mammals in the world: the duckbill platypus and the spiny anteater (echidna). See "Resources" (page 307) for books by Ruth Heller about these groups of mammals.*

7. Ask a few students to share their statements about what the organisms from the **Reptiles** sheet had in common with the organisms from the **Mammals** sheet. Alternate the **Reptiles** and **Mammals** transparencies on the projector during the discussion. Make sure these important points are considered:

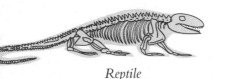

Reptile

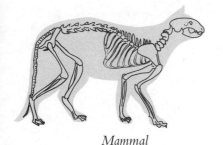

Mammal

What Reptiles and Mammals Have in Common

- **A backbone.** Their skeletons match up, bone type for bone type.

- **Watertight skin.**

- **A large intestine** for extracting water from waste products.

- **"One-way traffic" guts.**

---

8. Now that the class has witnessed the evolution of vertebrates from the first bony fish to the most recent mammals, show the overhead transparency **Skeletons through Time** on the projector. Ask a few students to point out the similarities in the skeletons of fish, amphibians, reptiles, birds, and mammals. [Fin bones are like limb bones, etc.]

9. Debrief the **Organism Adaptations** stations.

   Have students refer to the Organism Adaptations pages in their journal. Revisit each animal and plant from the stations, asking the students for a few of their ideas on each organism's adaptations.

10. Debrief the **"Guts"** station.

   a. Alternate the two overhead transparencies **"Guts": Single-Celled Organism/Sponge/Jellyfish/Earthworm** and **"Guts": Bird and Mammal.** Have students note the similarities and differences among these systems. During the discussion, point out that although mammals' digestive parts are more specialized (do specific jobs), they're quite similar to an earthworm's.

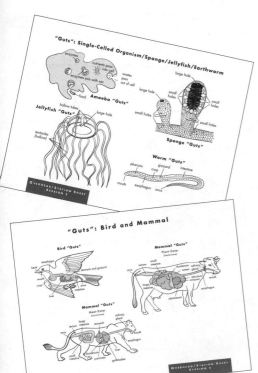

b. Add the station sheet **"Guts": Bird and Mammal** to the **Life through Time** wall chart.

11. Debrief the **Continental Drift** station.

a. Review how the continents appeared during the previous time period on the **Life through Time** wall chart and ask students how the continents moved during the time period they just studied. Be sure they raise the following points:

- South America has separated from Antarctica and Africa.
- North America has separated from Europe.
- India has moved north and slammed into Asia, forming the Himalayan range.
- Australia has separated from Antarctica and moved north.
- The Rockies and the Andes have formed.
- Land now links the Americas, and the migration of top predators from North America has wiped out many South American animals.
- Global climate has cooled; giant glaciers and ice caps have formed again in North America, Eurasia, and at the poles.
- By the end of this session, the continents have drifted into the positions we know today—and the drift continues!

b. Add the station sheet **Continental Drift—Time Period #5** to the **Life through Time** wall chart.

We've included two *views* of continental drift for this final time-travel session, so that students can picture ("Aha!") the geologic creep of land masses as they assembled into the continents we know today. Remind the class that the continents are still moving; what we think of as our fixed world is just a transient state of affairs in our planet's evolution.

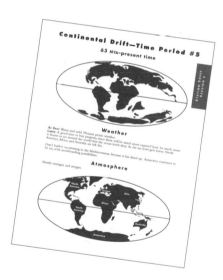

■ Major Evolutionary Events

1. Ask students to consider all they've just discussed, then brainstorm some major evolutionary changes that took place in the organisms or their habitats during the period they just studied, referring to the list below as a content checklist.

> **Major Evolutionary Events—Time Period #5**
>
> - Mammals are abundant; most huge mammals die off, grazing mammals increase.
>
> - First primates appear.
>
> - Grasses spread.
>
> - Continents have moved into their present-day locations.
>
> - The global climate has cooled.
>
> - In the very, very recent past, **the first humans evolve,** as primates venture from a forest lifestyle to the plains.

2. Ask students how their observations compare with the predictions they made in Session 5.

3. Post the sheet **Major Evolutionary Events—Time Period #5** above this session's strip of the **Class Time Line.**

4. Tell the students that in the next and final session, they'll have a chance to be inventive and express the concepts they've learned throughout the unit.

■ Most Representative Organism of the Age

1. Tell students it's time to name this time period.

2. Hand out the cut-up speeches to those who'll play each role.

3. Have the volunteers read their speeches, in numerical order and begin.

4. Reconvene the class and review the candidates:

- The Age of Birds
- The Age of Humans
- The Age of Mammals

5. Conduct the final election of the unit and tally the vote. Write the name of the winner on the **Age of**_____ sign, a blank sheet, or directly on the wall chart. Have students write it in their journals. If you made a sign, add it to the chart.

Note: Your **Life through Time** wall chart is now complete, and should look like this:

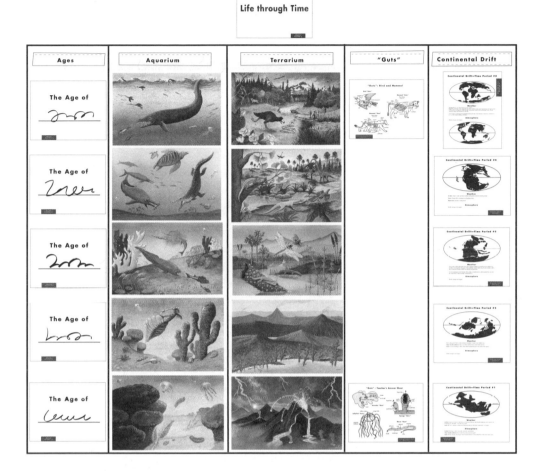

6. If you've read ahead and chosen the option you'll use in Session 7, you may want to ask students to bring in materials useful to whatever projects they'll be working on.

■ Going Further

1. Incorporating Related Activities

Any activities you have from other curricula relating to pollination, grasses, mammals, birds, the rise of consciousness, or human impact on the planet would be an excellent addition at this point.

2. Song

The songs *One Of These Days We're Gonna Rule The World (The Cockroach)* and *Family Tree* (pages 328 and 329) both provide overall review of general evolutionary trends. Hear audio clips at www.moo-boing.com.

Script for Most Representative Organism Election for
Time Period #5: 65 MYA–present time

1

Bird: Life started in the seas, moved onto land, and **finally** moved into the air. We **clearly** rule the air. No other animal with a backbone can fly (except bats). Humans can just dream about it, but we *do* it. This should be called the "Age of Birds."

2

Human: We can *too* fly. With our machines, we can go anywhere on the surface of the Earth, and even to other planets, beyond where any other Earth creature can. We may not always do a good job, but we **totally** rule the Earth. We've completely changed entire habitats, and killed off many species. We've made some discoveries that benefit the Earth and its life-forms, and many that are lousy for our planet. We've even begun to make new animals and plants through breeding and genetics. How could this not be called "the Age of Humans"?

3

Opossum: Excuse me, but this is a **65-million-year period of time** we're talking about here, and humans have only been around for **6 to 7 million years!** You just got here! Yes, I'd agree that you've been a significant species for the last million years or so, but you can't exactly be prevalent if you don't even **exist** for the other **60 million years** or so! But humans are part of a larger group that **has** been prevalent for this time—the mammals. And birds? Give me a break, they're just feathered reptiles, and far from prevalent in this period. This should be called the "Age of Mammals"!

Birds

Mammals

Skeletons through Time

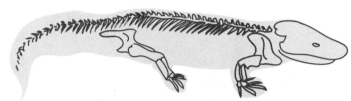
Lobe-finned Fish

Early Amphibian

Reptile

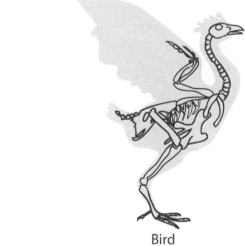
Bird

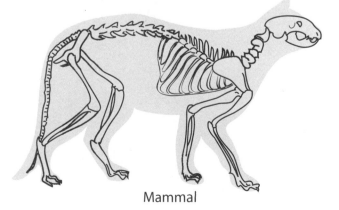
Mammal

"Guts": Bird and Mammal

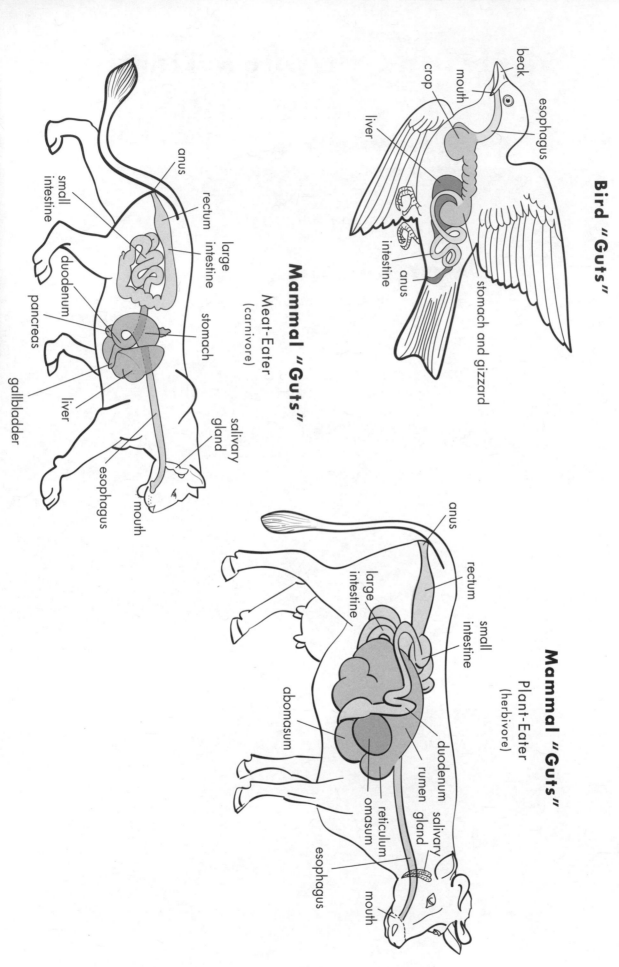

Bird "Guts"

beak
mouth
crop
esophagus
liver
small intestine
anus
stomach and gizzard

Mammal "Guts"
Meat-Eater
(carnivore)

anus
small intestine
rectum
large intestine
duodenum
pancreas
gallbladder
liver
stomach
esophagus
mouth
salivary gland

Mammal "Guts"
Plant-Eater
(herbivore)

anus
rectum
small intestine
large intestine
duodenum
salivary gland
rumen
reticulum
omasum
abomasum
esophagus
mouth

Major Evolutionary Events—Time Period #5

- Mammals are abundant; most huge mammals die off, grazing mammals increase.

- First primates appear.

- Grasses spread.

- Continents have moved into their present-day locations.

- The global climate has cooled.

- In the very, very recent past, **the first humans evolve,** as primates venture from a forest lifestyle to the plains.

Fossils—Time Period #5

© 2003 The Regents of the University of California. May be duplicated for classroom or workshop use.

Aquarium Background —Time Period #5 • *Session 6*

Terrarium Background —Time Period #5 • *Session 6*

Continental Drift—Time Period #5

65 MYA-present time

Weather

At first: Warm and mild. Pleasant picnic weather.
Later: A good time to buy property, since there will be much more exposed land. So much water is frozen in ice around the world that the ocean levels drop. As the sea level gets lower, North America, Africa, and Australia are left dry.

Don't bother vacationing in the Mediterranean, because it has dried up. Antarctica continues to be icy, with snowboarding possibilities.

Atmosphere

Mostly nitrogen and oxygen.

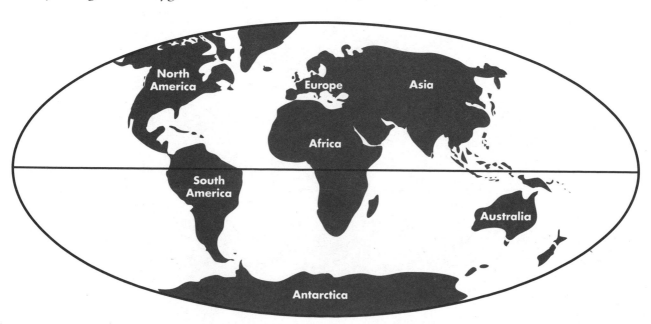

Time Travel Journal —Time Period #5

65 million years ago to present time

The Age of _____

Draw and label animals and plants you identify.

Time Travel Aquarium

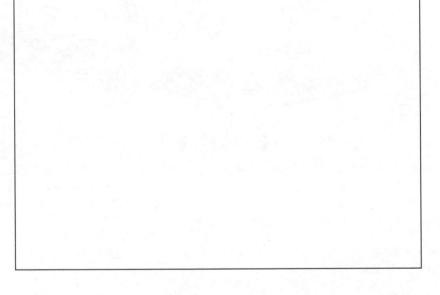

Time Travel Terrarium

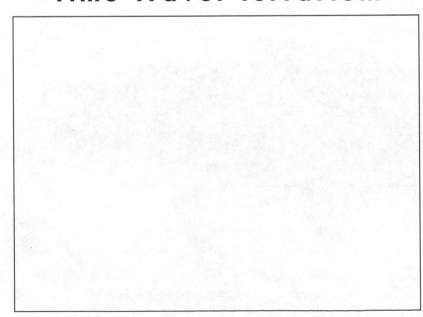

Fossils

Label any fossils you're able to identify.

Birds

Look at the pictures and write two or more statements about what these organisms have in common and how they are different.

_____

_____

_____

_____

_____

Mammals

Look at the pictures and write two or more statements about what these organisms have in common and how they are different.

_____

_____

_____

_____

_____

Organism Adaptations

Carefully observe the animals and plants, and write your best guesses to these questions:

Animal	What does it eat?	What eats it?	How does it move around?	How does it protect itself?

Plant	How does it feel?	What kind of environment does it live in? (Wet? Dry? Very dry?)	How does it reproduce?

"Guts"

Use a pencil to trace the path you think food
takes through the bird and mammals.

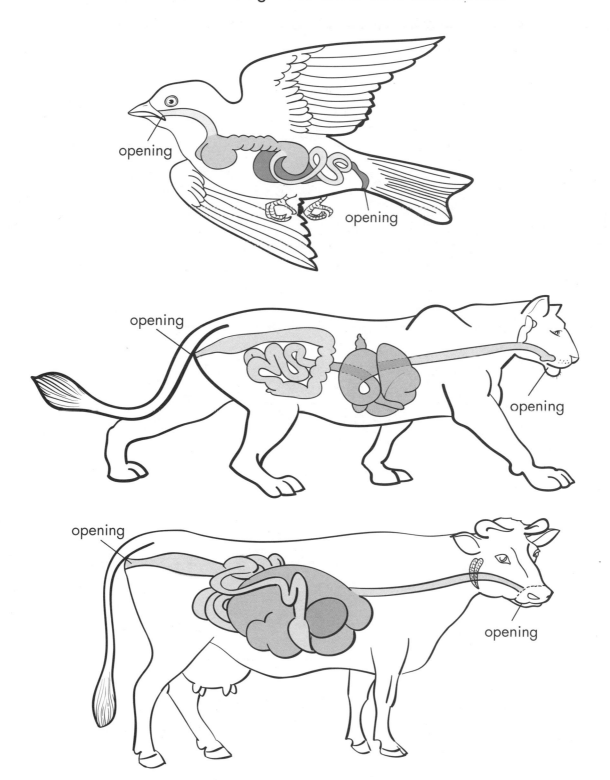

Continental Drift

Draw the continents as they appear today, referring to the station sheet. Describe the weather and atmosphere.

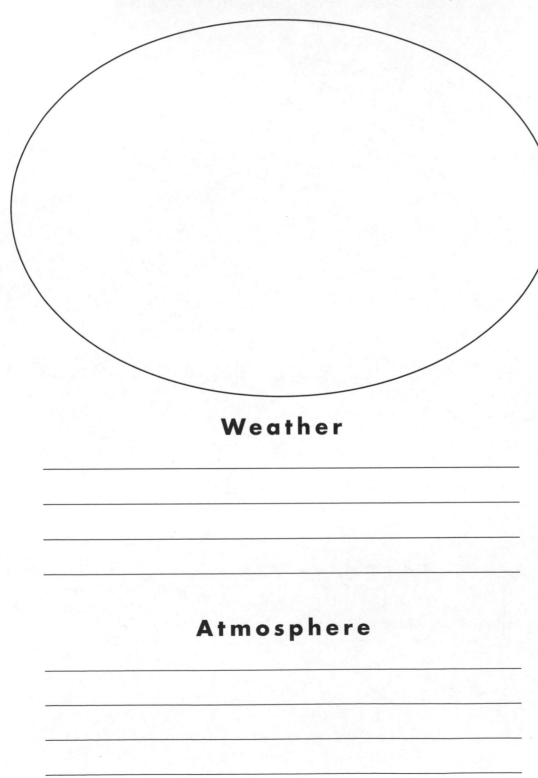

Weather

_____

_____

_____

_____

Atmosphere

_____

_____

_____

_____

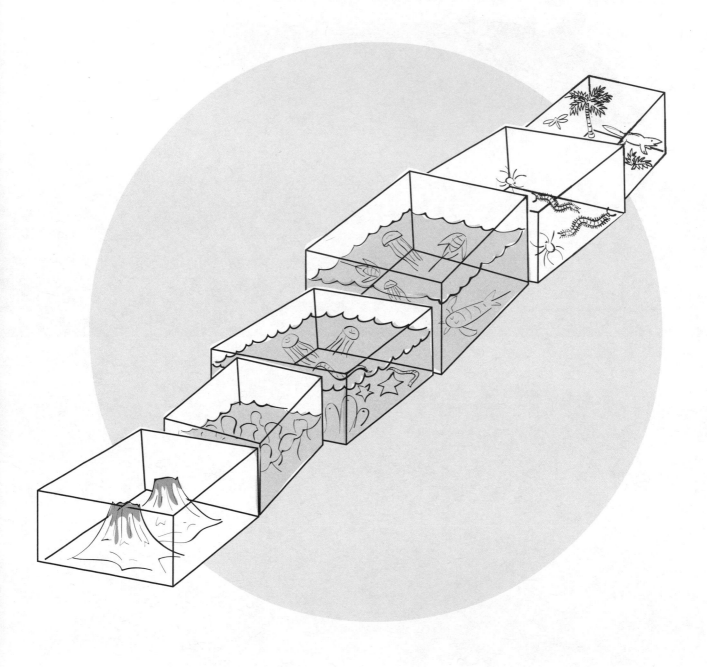

Overview

In this closing session, students are challenged to apply what they've learned throughout the unit in a creative project. There are several options for this wrap-up session. Depending on your energy level, your students' enthusiasm, and your time constraints, the choice of activities ranges from writing projects that can largely be done outside of class ("Time Traveler Adventure Stories" and "Dramatizing Life through Time") to classroom projects ("Explosions and Extinctions" and "3-D Class Diorama"). The first and last assignments use **Period Information Sheets** to flesh out the standard geologic time line that paleontologists and other scientists use. In the other two assignments, a brief class discussion on the standard time line is highly recommended, but optional.

Here are the assignment options:

1. Time Traveler Adventure Stories

After a whole-class discussion about the standard geologic time line that scientists use, students are divided into small groups. Using **Period Information Sheets** and referring to questions posted on the board, each group brainstorms what it would be like to travel to a past geologic period. The students are then assigned (in class or as homework) to write individual fictional—but scientifically accurate—stories of their adventures in the past, including as many elements of the time as possible: descriptions of the life-forms, continents, atmosphere, and weather. This option begins on page 236.

2. Dramatizing Life through Time

This option encourages students to produce songs, skits, poems, and other representations of how life on Earth has evolved over time. Much of the students' work can be assigned as homework, with an optional classroom presentation to follow. A brief class discussion on the standard time line is recommended, but optional. This option begins on page 240.

3. Explosions and Extinctions

This option challenges students to come up with explanations for one of the major evolutionary events in life on Earth, taken from the **Class Time Line.** They attempt to explain the extinction, decline, or population explosion of one or more species of animals and/or plants. They must support their argument with evidence based on the position of the continents, the weather and atmosphere, and the other animals and plants of the time. This can be a writing assignment, a mural with an oral description, or a skit. This option begins on page 243.

4. **3-D Class Diorama**

In this most elaborate option, the class is introduced to the standard geologic time line that paleontologists use to describe "deep time." Teams of students using **Period Information Sheets** create their own scenes of specific geologic time periods and finally put them all together to form a three-dimensional classroom model of life through time. This is a powerful visual representation of everything students have learned, and reinforces their sense of the evolution over time of life on our planet. This option begins on page 248.

■ What You Need for All Options

For the class:
- ❑ the completed **Class Time Line** from Session 6
- ❑ the completed **Life through Time** wall chart from Session 6
- ❑ chalkboard, chart paper, or a blank overhead transparency for listing questions and reminders for the class (see each "Getting Ready")
- ❑ *(optional)* mural paper and colored pens and pencils
- ❑ *(optional)* chart paper for making signs

For each student:
- ❑ **Time Travel Journal** from the previous session
- ❑ **Organism Key** from the previous session
- ❑ blank or lined paper on which to write the assignment

Option 1: Time Traveler Adventure Stories

■ Additional What You Need for Option 1

For the class:
- ❑ 1 overhead transparency of **Standard Geologic Time Line** (page 253)
- ❑ 1 overhead transparency of **Time Line Comparison Chart** (page 254)
- ❑ an overhead projector and screen

For each team of four students:

❑ 1 set of **Period Information Sheets** (pages 255–268; 14 pages total per team)

■ Getting Ready

1. Copy and set aside one set of **Period Information Sheets** (pages 255–268; 14 pages total) for each team of four.

2. Decide whether you'll have students do the assignment independently, in pairs, or in teams of four. (In all cases, students will begin in teams of four to share the **Period Information Sheets.**)

3. Write the following questions on the board, chart paper, or blank transparency:

   • What geologic time period would you like to travel to, and why?
   • What might the atmosphere and weather be like?
   • What might the land be like? the water?
   • What might the continents be like?
   • What kinds of animals and plants might you see?

4. Copy the overhead transparencies for this option: **Standard Geologic Time Line** (page 253) and **Time Line Comparison Chart** (page 254).

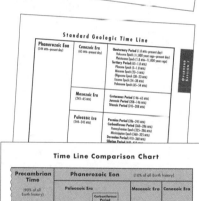

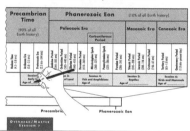

5. Referring to the "Ages" column on the **Life through Time** wall chart, write in the name of each session's Age on the **Time Line Comparison Chart** transparency.

■ Introducing the Standard Geologic Time Line

1. Tell the students that today they're going to write a fictional travel adventure story about a voyage to a period in Earth's history.

2. Explain that although they've been representing geologic time throughout the unit by naming the ages for themselves according to the most representative organism(s), other scientists represent the "deep time" of Earth's history with a **standard geologic time line** that can show all the billions of years of Earth's history at once.

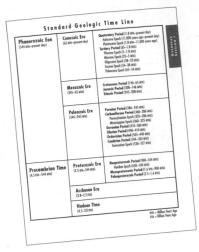

3. Show the **Standard Geologic Time Line** transparency on the overhead projector. Point out the different labels scientists use for various "chunks" of time:

- **Eon or Time**
  When the Earth's geologic history is divided up into only **two** huge chunks, they're called the Precambrian **Time** and the Phanerozoic **Eon.**
- **Era**
  Eon and time can be broken down into smaller chunks of time called **eras.**
- **Period**
  Eras can be broken down even further, into **periods.**
- **Epoch**
  Most periods are broken down into still smaller chunks, called **epochs.**

4. Show the **Time Line Comparison Chart** transparency. Take a moment with the class to review how the time periods they studied (the ages they named for themselves) break down into periods of geologic time devised by scientists.

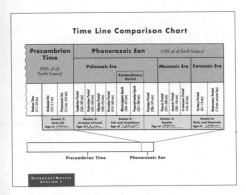

■ Introducing the Period Information Sheets

1. Divide the class into teams of four.

2. Tell students that each team will be given a set of **Period Information Sheets** to look through. Ask the students to imagine traveling back in time as you summarize the kind of information they're about to see on these sheets.

3. Explain that each student will be given time to write a fictional but realistic adventure story about one of the time periods. Unless you've decided to assign this option as a pair or team effort, let students know that each one of them will be allowed to choose the period she wants to travel to. (It doesn't matter if other students in the team choose the same period.)

4. Hand out the **Period Information Sheets,** one set per team.

5. Give the teams a few minutes to review the sheets at their tables.

6. Using one time period as an example, conduct a class brainstorm about each of the questions you wrote on the board. Be sure the

students' responses include some ideas about how each of the conditions might affect them if they lived in that time.

- *What geologic time period would you like to travel to, and why?*
- *What might the atmosphere and weather be like?*
- *What might the land be like? the water?*
- *What might the continents be like?*
- *What kinds of animals and plants might you see?*

■ Setting Up the Assignment

1. Remind the students that each of them will be writing a travel adventure story about a period in Earth's past, referring to her **Time Travel Journal** and **Organism Key,** and the **Class Time Line.** Point out that it helps readers become involved in the story when writers describe in detail what they see, hear, feel, taste, and smell. Say that the students may first want to set the scene, describing their surroundings, before writing about their adventures and experiences. They may illustrate these stories if they wish.

2. Explain that although the idea of time travel itself is not realistic, they'll need to keep their stories realistic—no talking dinosaurs or alien spaceships!

3. Describe the criteria you'll be using to assess or grade their stories. Let them know their stories should include as much information as possible on their time period's atmosphere and weather, continents, land, water, animals, and plants.

4. Distribute paper to all students and have them write on it their names, the title **Time Traveler Adventure Story,** and the name of the time period they've chosen to write about.

5. Let students know how much time they'll have to complete the assignment, and let them begin. You may choose to let students do or finish this assignment as homework, in which case they can also do research on their own using reference books or the Internet.

6. Depending on time, ask a few students to share their stories with the class, or post the stories in the classroom where students can read them.

7. If you wish, add these stories to the students' journals as part of their "portfolios" for assessment.

Option 2: Dramatizing Life through Time

■ Additional What You Need for Option 2

optional but recommended:
For the class:
- ❏ 1 overhead transparency of **Standard Geologic Time Line** (page 253)
- ❏ 1 overhead transparency of **Time Line Comparison Chart** (page 254)
- ❏ an overhead projector and screen

■ Getting Ready

1. Write this "starter list" of ideas on the board:

    a. Acting the time line
    b. Singing the time line
    c. Reciting the time line
    d. Remembering the time line
    e. ?

2. Decide whether you'll have students work on this assignment independently or in pairs or larger groups.

3. *If you've decided to introduce the standard geologic time line:*

    a. Copy the overhead transparencies for this option: **Standard Geologic Time Line** (page 253) and **Time Line Comparison Chart** (page 254).

    b. Referring to the "Ages" column on the **Life through Time** wall chart, write in the name of each session's Age on the **Time Line Comparison Chart** transparency.

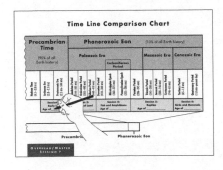

Optional but Recommended:

■ Introducing the Standard Geologic Time Line

1. Tell the students that today they'll be asked to choose a way to **dramatize** the **Class Time Line.**

2. Explain that although they've been representing geologic time throughout the unit by naming the ages for themselves according to the most representative organism(s), other scientists represent the "deep time" of Earth's history with a **standard geologic time line** that can show all the billions of years of Earth's history at once.

3. Show the **Standard Geologic Time Line** transparency on the overhead projector. Point out the different labels scientists use for various "chunks" of time:

   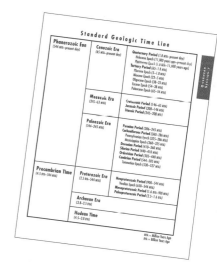

   - **Eon or Time**
     When the Earth's geologic history is divided up into only **two** huge chunks, they're called the Precambrian **Time** and the Phanerozoic **Eon.**
   - **Era**
     Eon and time can be broken down into smaller chunks of time called **eras.**
   - **Period**
     Eras can be broken down even further, into **periods.**
   - **Epoch**
     Most periods are broken down into still smaller chunks, called **epochs.**

4. Show the **Time Line Comparison Chart** transparency. Take a moment with the class to review how the time periods they studied (the ages they named for themselves) break down into periods of geologic time devised by scientists.

■ Setting Up the Assignment

1. Tell (or remind) the students that today they'll be asked to choose a way to dramatize the **Class Time Line.** They'll be free to refer to any of the charts, dioramas, or stations available in the classroom. They can also use their **Time Travel Journals** and **Organism Keys** from previous sessions.

2. If you've decided to assign this option as homework, encourage students to conduct research on their own, using books or the Internet.

3. Explain that you've prepared a "starter list" for them: three or four possible ways in which they can interpret and share what they've learned about life through time. Emphasize that they're not limited to this list and should use their imaginations!

4. Let students know they'll conclude this assignment by making presentations to the class, if you've decided to do this.

5. Define your expectations for the students' assignments. Say that their presentations should show that they understand the concept that plants, animals, weather, atmosphere, land, water, and continental distribution have changed (evolved) over billions of years, and that their assignments should include factual information about the different periods.

6. Show students the starter list you wrote on the board and briefly explain each option:

   a. **Acting the time line**
      Students script (and optionally perform) a short play to dramatize the evolution of life through time.

   b. **Singing the time line**
      Students write (and optionally perform) a song that describes the evolution of life through time.

   c. **Reciting the time line**
      Students write (and optionally recite) a poem that describes the evolution of life through time.

   d. **Remembering the time line**
      Students assemble a sentence or other mnemonic—along the lines of the colors-of-the-rainbow memory trick "**R**ed **O**range **Y**ellow **G**reen **B**lue **I**ndigo **V**iolet = **Roy G. Biv**"—for remembering the names and order of the individual *eras, periods,* and *epochs* from the geologic time line.

   e. **?**
      Remind students that they're not limited to this list, and should feel free to use other dramatic ways (perhaps mimes or charades) to represent life through time, as long as they meet the requirements you described earlier.

7. If you've decided to have students work in pairs or groups, divide the class now. If you've decided to assign part or all of this option as individual homework, let students know that their presentations will be made when the class next meets.

8. Distribute paper to all students and have them write on it their names and the title **Dramatizing Life through Time.** If they'd like the reminder, they can also write down the starter-list of questions you posted.

9. Have the class begin or take their assignments home.

10. If you've decided to do this, have each student or group make a presentation to the class.

Option 3: Explosions and Extinctions

■ Additional What You Need for Option 3

For the class:
- ❑ 1 copy of **Explaining Major Evolutionary Change** (page 269)

optional but recommended:
- ❑ 1 overhead transparency of **Standard Geologic Time Line** (page 253)
- ❑ 1 overhead transparency of **Time Line Comparison Chart** (page 254)
- ❑ an overhead projector and screen

■ Getting Ready

1. Copy, enlarge, and post the list called **Explaining Major Evolutionary Change** (page 269), or write it on the board or chart paper.

2. *If you've decided to introduce the standard geologic time line:*

   a. Copy the overhead transparencies for this option: **Standard Geologic Time Line** (page 253) and **Time Line Comparison Chart** (page 254).

   b. Referring to the "Ages" column on the **Life through Time** wall chart, write in the name of each session's Age on the **Time Line Comparison Chart** transparency.

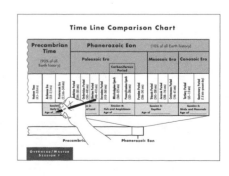

Optional but Recommended:

■ Introducing the Standard Geologic Time Line

1. Tell the students that today they get to use the **Class Time Line** they completed in Session 6 to answer one or more major questions they'll generate about evolutionary change.

2. Explain that although they've been representing geologic time throughout the unit by naming the ages for themselves according to the most representative organism(s), other scientists represent the "deep time" of Earth's history with a **standard geologic time line** that can show all the billions of years of Earth's history at once.

3. Show the **Standard Geologic Time Line** transparency on the overhead projector. Point out the different labels scientists use for various "chunks" of time:

- **Eon or Time**
  When the Earth's geologic history is divided up into only **two** huge chunks, they're called the Precambrian **Time** and the Phanerozoic **Eon.**
- **Era**
  Eon and time can be broken down into smaller chunks of time called **eras.**
- **Period**
  Eras can be broken down even further, into **periods.**
- **Epoch**
  Most periods are broken down into still smaller chunks, called **epochs.**

4. Show the **Time Line Comparison Chart** transparency. Take a moment with the class to review how the time periods they studied (the ages they named for themselves) break down into periods of geologic time devised by scientists.

■ Setting Up the Assignment

1. Tell (or remind) your students that today they get to use the **Class Time Line** they completed in Session 6 to answer major questions about evolutionary change. Their assignment is to come up with one or more questions about major changes in life on Earth, and then write or present arguments for what they think caused the change(s) to occur.

2. Remind students that in studying fossils, paleontologists have noticed certain patterns of change in organisms. There are long periods of little change, followed by sudden bursts, or "punctuations," of significant change. These bursts seem to occur when the environment changes; either because something new is added to the system, or because the environment has deteriorated.

3. Referring to the **Class Time Line** on the wall, briefly review with the students how sometimes a huge group of animals becomes extinct, sometimes a group flourishes and takes over, and sometimes a group that was once prevalent declines but still survives. Paleontologists have attempted to come up with explanations for these patterns—and now the students will do the same!

4. Tell them that as an example of their assignment you're going to run through a sample question with the class—it's the best-known of these types of questions: "Why did the dinosaurs die off?"

■ Taking the Class through a Sample Question

1. First, ask the class to come up with **EVIDENCE** that the dinosaurs did die off, using the **Class Time Line** and/or the **Life through Time** wall chart. [They should mention the large numbers of dinosaur fossils dating from the Permian through the Cretaceous Periods (Session 5), and the lack of fossils after that point.]

2. Next, ask the class to consider this: **How might the shifting of CONTINENTS affect organisms?**

   a. Draw your students' attention to the Continental Drift column on the **Life through Time** wall chart. Point out the changes that the continents went through before, during, and at the end of the dinosaurs' reign.

   b. Have students think of some ways in which the positioning of the continents might have affected the dinosaurs. Bring up the following possibilities, if not mentioned by your students:

   • **Change in temperature:** A continent shifting closer to or farther from the equator would become hotter or colder. How might this affect organisms on Earth? [Animals and plants not able to adapt to the new temperature might die off, while animals that could adapt might flourish.]

- **New land masses:** What might happen to plant and animal life if continents and islands that had been separated for huge periods of time by large seas became connected? [These life-forms would invade each other's habitat. Some animals and plants could out-compete others and flourish, while the others died off.]

- **Changing habitat:** What effect might it have on animals and plants adapted to a specific habitat, such as large shallow seas, low-lying plains, or mountains, if their habitat were destroyed by shifting continents? [Other animals, better able to adapt to the new habitat, could flourish.]

3. Now have the class think about this: **How might changes in a PLANT OR ANIMAL affect other organisms?**

Draw the students' attention to the **background illustrations** on the **Life through Time** wall chart. Have them think of ways in which changes in plant or animal life before, during, and at the end of the dinosaurs' reign might have affected the dinosaurs and other organisms. Be sure the following possibilities are brought up:

- **New kinds of plants:** What effect might a newly evolved type of plant, such as the first land plants or cone-bearing plants, have on other organisms? [It would provide a huge new food source. Animals that were able to eat it could flourish, while many of their former competitors died off.]

- **New kinds of animals:** What if a type of animal evolved to eat a new food source (a new type of plant or animal)? How might that affect other animals? [That new animal would itself provide a new food source for other animals. When grasses evolved, for example, deer-like animals and other grass eaters became a new kind of prey for meat eaters. Animals that could eat these new grazers flourished, while competitors that couldn't died off.]

- **New strategies or features:** If an animal or plant evolved a new strategy or characteristic—such as the ability to photosynthesize, or an egg and skin that hold in water, or wings for flight—what effect might that have on other organisms? [It could out-compete other animals or plants.]

- **Species interdependence:** What might happen if a species evolved to be very dependent on another species, and the species it depended on died off? [It would be vulnerable to extinction. For example, a flowering plant that depends on only one insect species to pollinate it could be doomed to die off if its pollinator becomes extinct.]

4. Now ask the class, **How might a CATASTROPHIC EVENT affect organisms?**

Have your students name some catastrophic events that might affect life on Earth and explain *how* the organisms might be affected. Bring up the following possibilities if your students don't:

  • **Change in global temperatures.** If the planet's overall temperature rises or falls, animals and plants adapted to specific temperatures could decrease or increase in number or prominence.

  • **Volcanic eruptions** can destroy local habitat and/or species of animals and plants. Massive volcanic eruptions might cause enough dust in the atmosphere to lower global temperatures.

  • **An asteroid or comet collision** could also destroy local habitat and/or species, and might also lower global temperatures if excessive dust in the atmosphere blocked the Sun.

  • **Hurricanes** can destroy local habitat and/or species.

  • **Toxic gases** can occur in lethal quantities. Oxygen did, after the first photosynthesizing plants evolved, and proved deadly to early life. In some parts of the world, pockets of gas (methane hydrates) seep up from the Earth or are struck during drilling. This gas can take the place of oxygen in the atmosphere and kill organisms in the area.

5. Finally, ask the students what kind of **EVIDENCE** they'll need to support their explanations. Note that it's possible to come up with an explanation even if they have no direct evidence, provided they acknowledge that fact and describe what evidence they would need to verify their hypothesis.

■ Starting Students on Their Own Questions

1. Remind your students that their assignment is to come up with one or more questions about major changes in life on Earth, and then write or present arguments for what they think caused the change(s) to occur.

2. Call students' attention to the list you posted called **Explaining Major Evolutionary Change.**

3. Describe to the class any other specifics about the assignment, such as whether you want them to work in teams or individually, length of written assignment or presentation, and options for types of presentation (written paper, mural, skit, etc.).

4. Distribute paper to all students and have them write on it their names, the title **Explosions and Extinctions,** and the question(s) about major change that they'll be posing and answering. Depending on time, you may choose to have them finish their assignments at home.

5. Have a few students share their presentations with the class. For students who've done a written assignment, ask a few to read their papers, or post them on the wall for the class to read.

Option 4: 3-D Class Diorama

■ Overview

Having worked through the unit by naming the ages for themselves according to the most representative organism(s), students will now be working with the **standard geologic time line** that paleontologists use, the scientific "language" of evolutionary time.

You can opt to have students make just a two-dimensional drawing or collage, but we highly recommend the 3-D format.

Each team of two or three students is given a **Period Information Sheet** and challenged to make a diorama of the period described on that sheet. They're encouraged to be creative and use plastic models, drawings (from the information sheet or their own sources), and even (if you're OK with it) live organisms from the time-travel organism adaptations stations or others they bring in. When the students have finished, the dioramas are assembled in chronological order and displayed in the classroom to create a 3-D history of life on Earth from its first appearance to the present day.

■ Additional What You Need for Option 4

For the class:

- ❑ 1 overhead transparency of **Standard Geologic Time Line** (page 253)
- ❑ 1 overhead transparency of **Time Line Comparison Chart** (page 254)
- ❑ 1 paper copy **Time Line Comparison Chart** (page 254) to hang in the classroom
- ❑ an overhead projector and screen
- ❑ additional materials with which to make the dioramas: poster paper, scissors, reference books, dirt, plants, plastic animals and plants from the Time Travel Aquarium and Terrarium

For each team of two to three students:

- ❑ a copy of the **Period Information Sheet** for the period they'll be working on (pages 255–268)
- ❑ *(optional)* a shoebox or other small container (plastic, if any of the dioramas will be aquatic!) to be used for the diorama (Students can bring these containers in; see note on page 208 in Session 6.)

■ Getting Ready

1. Make a copy of each **Period Information Sheet** (pages 255–268), and choose a geologic time period to assign to each team of two to three students:

There are 14 geologic time-spans for this activity: 10 periods and two epochs plus the Archaean and Proterozoic Eras; this gives each two-student team in a class of 28 its own **Period Information Sheet** to work on. If you have fewer or more students, you can combine early periods (when there wasn't as much diversity of life-forms) or assign some teams of three. (See also "More Class-Size Alternatives," in margin.)

Phanerozoic Eon

    1. Quaternary Period: 1 million, 800,000 years ago–present day
    2. Tertiary Period: 65 million–1 million, 800,000 years ago
    3. Cretaceous Period: 146 million–65 million years ago
    4. Jurassic Period: 208 million–146 million years ago
    5. Triassic Period: 245 million–208 million years ago
    6. Permian Period: 286 million–245 million years ago
    7. Pennsylvanian Epoch: 325 million–286 million years ago
    8. Mississippian Epoch: 360 million–325 million years ago
    9. Devonian Period: 410 million–360 million years ago

More Class-Size Alternatives

1. For one less student group: Time-spans #7 and #8 (the Pennsylvanian and Mississippian Epochs) can be combined into the **Carboniferous Period.** The team is given the **Period Information Sheets** for both epochs.

2. For one additional student group: The **Hadean Era,** the first 700 million years of the planet's existence, was a cataclysmic but lifeless period on Earth; full of atmospheric and geologic sound and fury, but more challenging for students to "travel" to. If you need to provide another team period, however, the Hadean (4.5–3.8 BYA) could conceivably be an option, with emphasis on the molten, gaseous, cosmic circumstances of the planet's physical formation. There's no **Period Information Sheet** for the Hadean, but the Internet is a good source of information on this era.

Each time-span listed here belongs within one of four eras: The Proterozoic, Paleozoic, Mesozoic, or Cenozoic. See the **Standard Geologic Time Line** *(page 253) for the big-picture breakdown.*

10. **Silurian Period:** 440 million–410 million years ago
11. **Ordovician Period:** 505 million–440 million years ago
12. **Cambrian Period:** 544 million–505 million years ago

Precambrian Time

13. **Proterozoic Era:** 2 billion, 500 million–544 million years ago (contains three periods)
14. **Archaean Era:** 3 billion, 800 million–2 billion, 500 million years ago

2. Designate an area of the classroom in which to assemble the 3-D Class Diorama when it's time to put the individual dioramas together. This could be a countertop, desks and tables pushed together, or the floor. (If you assign each team of two a 12-inch x 6-inch space, for instance, you'll need a 14-foot-long area for the assembled diorama.) It could also be wall space, if you specify to your students that their dioramas must be light enough to tack or tape to the wall—no organisms, of course!

3. Decide whether you'll allow your students to include live animals in their dioramas, and what your limitations will be (including what animals would and wouldn't be OK to use, how to collect them, how to contain them, etc.).

4. Copy the overhead transparencies for this option: **Standard Geologic Time Line** (page 253) and **Time Line Comparison Chart** (page 254).

5. Referring to "Ages" column on the **Life through Time** wall chart, write in the name of each session's Age on the **Time Line Comparison Chart** transparency and paper copy.

6. Hang the paper copy of the **Time Line Comparison Chart** next to the **Life through Time** wall chart or elsewhere on the wall.

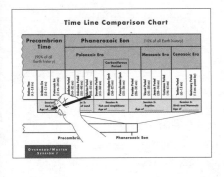

GO ■ **Introducing the Geologic Time Line**

1. Tell the students that today, in teams of two or three, they'll begin making a three-dimensional ("3-D") Class Diorama of the history of life on Earth.

2. Explain that although they've been representing geologic time throughout the unit by naming the ages for themselves according to the most representative organism(s), other scientists represent the

"deep time" of Earth's history with a **standard geologic time line** that can show all the billions of years of Earth's history at once. The students are going to be using that time line today.

3. Show the **Standard Geologic Time Line** transparency on the overhead projector. Point out the different labels scientists use for various "chunks" of time:

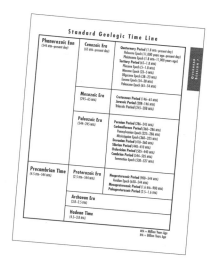

- **Eon or Time**
  When the Earth's geologic history is divided up into only **two** huge chunks, they're called the Precambrian **Time** and the Phanerozoic **Eon.**

- **Era**
  Eon and time can be broken down into smaller chunks of time called **eras.**

- **Period**
  Eras can be broken down even further, into **periods.**

- **Epoch**
  Most periods are broken down into still smaller chunks, called **epochs.**

4. Show the **Time Line Comparison Chart** transparency. Take a moment with the class to review how the time periods they studied (the ages they named for themselves) break down into periods of geologic time devised by scientists.

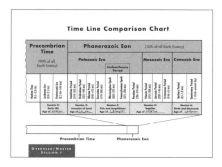

5. Show the class the same **Time Line Comparison Chart** you posted on the wall, and tell them it'll stay up for the duration of the assignment.

■ Setting Up the Assignment

1. Remind the students that although they've been focusing on time periods based on the evolution of major groups of animals and plants (the Age of _____), the 3-D Class Diorama they're going to assemble will be divided up into the *standard geologic periods,* as shown on the chart.

2. Explain that each team of students will be assigned a geologic time period and receive a **Period Information Sheet** about it. The sheet will show the animals and plants that lived at that time, the way the continents were positioned, and the state of the atmosphere and weather.

3. Ask the students to imagine traveling back in time as you summarize the kind of information they're about to see on these sheets.

4. Say that each team will create a model, or **diorama,** representing the period they've been assigned. Let them know that the team dioramas can include any of the materials you've provided at the front of the class—dirt, plants, plastic animals and plants from the Time Travel Aquarium and Terrarium, etc.—and can include paper cut-outs and/ or paper backgrounds. If you're willing to let them also include live animals from the organism adaptations stations or elsewhere, let them know the limitations and how to contain the organisms.

5. Let the class know that when all the teams' dioramas are completed, they'll be assembled and displayed in chronological order, creating a three-dimensional time line of the history of life on Earth.

6. Divide the students into teams of two or three, distribute the assigned **Period Information Sheets,** and let them begin.

■ Wrap-Up

1. When the students have completed their dioramas, ask them to line them up in chronological order in the classroom space you set aside for this purpose. Help them secure their dioramas into one continuous display.

2. Allow some time for students to tour and appreciate the completed diorama. (You may wish to keep it assembled if and while you teach relevant curricular material.)

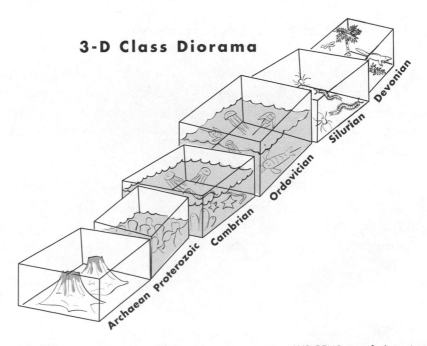

3-D Class Diorama

Archaean Proterozoic Cambrian Ordovician Silurian Devonian

Standard Geologic Time Line

Phanerozoic Eon
(544 MYA–present day)
	Cenozoic Era
	(65 MYA–present day)
		Quaternary Period (1.8 MYA–present day)
		   Holocene Epoch (11,000 years ago–present day)
		   Pleistocene Epoch (1.8 MYA–11,000 years ago)
		Tertiary Period (65–1.8 MYA)
		   Pliocene Epoch (5–1.8 MYA)
		   Miocene Epoch (23–5 MYA)
		   Oligocene Epoch (38–23 MYA)
		   Eocene Epoch (54–38 MYA)
		   Paleocene Epoch (65–54 MYA)

	Mesozoic Era (245–65 MYA)	**Cretaceous Period** (146–65 MYA) **Jurassic Period** (208–146 MYA) **Triassic Period** (245–208 MYA)

	Paleozoic Era
	(544–245 MYA)
		Permian Period (286–245 MYA)
		Carboniferous Period (360–286 MYA)
		   Pennsylvanian Epoch (325–286 MYA)
		   Mississippian Epoch (360–325 MYA)
		Devonian Period (410–360 MYA)
		Silurian Period (440–410 MYA)
		Ordovician Period (505–440 MYA)
		Cambrian Period (544–505 MYA)
		   Tommotian Epoch (530–527 MYA)

Precambrian Time (4.5 BYA–544 MYA)	**Proterozoic Era** (2.5 BYA–544 MYA)	**Neoproterozoic Period** (900–544 MYA) Vendian Epoch (650–544 MYA) **Mesoproterozoic Period** (1.6 BYA–900 MYA) **Paleoproterozoic Period** (2.5–1.6 BYA)
	Archaean Era (3.8–2.5 BYA)	
	Hadean Time (4.5–3.8 BYA)	

MYA = Million Years Ago
BYA = Billion Years Ago

Time Line Comparison Chart

		Hadean Time (4.5–3.8 BYA)		**Precambrian Time** (90% of all Earth history)
Age of Early Life	**Session 2:**	**Archaean Era** (3.8–2.5 BYA)		
		Proterozoic Era (2.5 BYA–544 MYA)		
Age of Invasion of Land	**Session 3:**	**Cambrian Period** (544–505 MYA)	**Paleozoic Era**	**Phanerozoic Eon** (10% of all Earth history)
		Ordovician Period (505–440 MYA)		
		Silurian Period (440–410 MYA)		
Age of Fish and Amphibians	**Session 4:**	**Devonian Period** (410–360 MYA)		
		Mississippian Epoch (360–325 MYA)	**Carboniferous Period**	
		Pennsylvanian Epoch (325–286 MYA)		
Age of Reptiles	**Session 5:**	**Permian Period** (286–245 MYA)		
		Triassic Period (245–208 MYA)	**Mesozoic Era**	
		Jurassic Period (208–146 MYA)		
		Cretaceous Period (146–65 MYA)		
Age of Birds and Mammals	**Session 6:**	**Tertiary Period** (65–1.8 MYA)	**Cenozoic Era**	
		Quaternary Period (1.8 MYA–present day)		

Precambrian Time

Phanerozoic Eon

Archaean Era

3 billion, 800 million years ago–2 billion, 500 million years ago
3.8–2.5 BYA

Life-Forms

The first life-forms on Earth appear in the Archaean Era. These are **bacteria,** in the seas; they'll be the only organisms on the planet for over one billion years. **Stromatolites,** mounds formed by colonies of photosynthesizing bacteria called **cyanobacteria,** or **blue-green algae,** carpet the landscape in shallow water. Because these structures trap sediment and secrete minerals, they will provide the oldest fossils on Earth. (Stromatolites still exist in modern times, but are uncommon.) Bacteria will turn out to be the single most successful organism on the planet, right to the present day. (There is even an ongoing debate about possible bacteria fossils found on Mars!)

Continental Drift

During the Archaean, the Earth's crust cools enough that rocks and continental plates begin to form. The oldest rocks found on Earth date to the beginning of the Archaean, 3 billion, 800 million years ago.

Atmosphere and Weather

The atmosphere is made up of methane, ammonia, hydrogen, carbon monoxide, and other gases that would be poisonous to most of the life on Earth today. Because cyanobacteria can photosynthesize, they begin to emit oxygen into the atmosphere; this will have a serious impact on life in the next era.

It's hot, with many volcanoes that send a lot of volcanic dust into the skies. There's thunder, lightning, and constant rain. Even the seas are hot.

Proterozoic Era

2 billion, 500 million years ago–544 million years ago
2.5 BYA–544 MYA

Life-Forms

Stromatolites, mounds of **cyanobacteria (blue-green algae)** that can photosynthesize, are still common in shallow waters at the beginning of the period but become more rare toward the end. (They're still around in modern times, but are not common.) There are many single-celled life-forms, including **archaea,** organisms that can withstand extreme environments like superheated or extra-salty water. (Some archaea are still found in hot sulfur pools, such as in Yellowstone.) At the end of the Proterozoic, significant life-forms called **eukaryotes** appear, the first organisms whose cells have a nucleus. (These are the ancestors of all animals, plants, fungi, and microorganisms we know today.) The **first multi-celled algae** and **multi-celled animals** appear at the end of the Proterozoic. All these organisms live in the sea, and they will provide a big burst of fossils to the geologic record.

Extinction: During the middle of the Proterozoic, there is a global catastrophe as accumulating oxygen from the photosynthesizing cyanobacteria saturates the atmosphere, killing off many other kinds of **bacteria.**

Continental Drift

The first stable continents have formed. The supercontinent Rodinia forms, centered at the South Pole.

Atmosphere and Weather

By the middle of this era, oxygen has built up substantially in the atmosphere. At least twice during the late Proterozoic the planet goes into a big freeze, with the polar icecaps a kilometer (more than half a mile) deep and extending to the equator. At these times Earth is very much like a giant snowball.

Cambrian Period

544 million years ago–505 million years ago
544–505 MYA

Life-Forms

So many new forms of life appear in the Cambrian Period that the phenomenon is often called the "Cambrian Explosion." **Early jellyfish, sponges, echinoderms** (starfish-like organisms), and **segmented worms** all appear at this time. (All these organisms have relatives still alive in modern times.) These new life-forms have evolved brand-new strategies for survival—actively hunting, tunneling deep into sediment, making complex burrows in the seabed. Tiny, spiny *Wiwaxia* creeps along the ocean floor, and **brachiopods** (clam-like animals) feed in very cold waters using hollow feeding tentacles.

Other kinds of animals are able to take chemicals from the water to make hard shells to protect themselves. These are the **arthropods,** and by far the most numerous are the **trilobites.** There are also **sea scorpions** and even a **lancelet**—the ancestor of the jawless fishes. All organisms live in the seas that cover most of the planet.

Continental Drift

The supercontinent Rodinia, which was centered at the South Pole, breaks apart. One of the "pieces," at the equator, is what will eventually become North America.

Weather

The weather is warming up as land fragments move toward the equator. Melting ice sheets create warm, shallow seas.

Ordovician Period

505 million years ago–440 million years ago
505–440 MYA

Life-Forms

Animals and plants still live in the sea, but there is evidence that **early plants** first start to invade the land during the Ordovician Period. There are many **trilobites,** and the **jawless fish** increase.

Mollusks appear: **bivalves,** with their two shells, look like modern clams and oysters. **Gastropods** have only one shell, and look like today's whelks and limpets. **Nautiluses, early squids,** and **octopuses** also come onto the scene.

Extinction: At the end of the Ordovician, when the shallow seas are drained as gigantic glaciers freeze the planet's waters, 60 percent of all ocean **invertebrates** go extinct.

Continental Drift

Most of the world's land is coming together to form the supercontinent called Gondwana. During the Ordovician, Gondwana moves toward the South Pole. Much of the continent, including what will one day be North America, is under water.

At the end of the Ordovician, Gondwana has settled at the South Pole and is covered by massive glaciers. With so much water frozen into glaciers, the shallow seas dry up and the worldwide sea levels drop.

Weather

During the early to middle Ordovician the weather is warm and the air is moist. At the end of the Ordovician, when continents are centered at the South Pole, massive glaciers form. The weather on the land gets much colder.

Silurian Period

440 million years ago–410 million years ago
440–410 MYA

Life-Forms

Trilobites, sea scorpions, mollusks, graptolites, conodonts, and **stromatoporoids** (sponge-like animals) are common. **Crinoids** (sea lilies) and **brachiopods** go through an explosive increase. **Jawless fish** continue to spread, and **freshwater fish** and **fish with jaws** first appear. **Coral reefs** form from the skeletons of tiny polyps.

There is clear evidence that life appears on land in the Silurian—organisms like **plants, fungi,** and **arthropod relatives of spiders and centipedes.**

Continental Drift

During the Silurian most continents are located in the Southern Hemisphere. Gondwana (what will be Australia, Antarctica, India, Africa, and South America) has settled around the South Pole. Over the course of this period the continents move north toward the equator and start to crash together to form one supercontinent called Pangaea.

Weather

During the Silurian, the weather becomes stable. Glaciers melt, causing the ocean to rise and form many warm, shallow sea basins.

Devonian Period

410 million years ago–360 million years ago
410–360 MYA

Life-Forms

At the beginning of the Devonian, the tallest plants are only a meter (just over three feet) tall. In the seas there are **sea scorpions, ammonites, corals, crinoids** (sea lilies), **starfish, sea urchins,** and **brachiopods.** There are still many **jawless fish.** Many new kinds of fish with jaws evolve, including **armored fish, sharks,** and **early bony fish.**

By the end of the Devonian there are **giant ferns, giant horsetails, club mosses,** and **seed plants,** which produced the **first trees** and the **first forests.** This huge increase in plants is sometimes called the "Devonian Explosion."

Amphibians move onto the land. The first terrestrial **arthropods** appear, including **early wingless insects** such as **silverfish** and **springtails. Early arachnids—spiders, ticks, mites,** and **scorpions**—increase. **Early millipedes** and **centipedes** probably appear during the Devonian.

Trilobites and **graptolites** become more rare. Most trilobites disappear by the end of the Devonian.

Continental Drift

During the Devonian, two supercontinents, Gondwana in the Southern Hemisphere and Laurentia to the north, are moving slowly toward each other. Gondwana (what will be Australia, Antarctica, India, Africa, and South America) and Laurentia (North America and northern Europe) are on a collision course.

Weather

Much of the land is under shallow seas. The weather is warm.

Mississippian Epoch
(also called the Lower Carboniferous)

360 million years ago–325 million years ago
360–325 MYA

Life-Forms

Snails, centipedes, scorpions and lots of kinds of **cockroach** increase on land. **Amphibians** diversify and spread. There are **giant horsetails, tree ferns, cycads, liverworts,** and **club mosses.** Lots of swamps. **Trilobites** are rare.

In the sea there are **sea scorpions, bryozoans, foraminifers,** and **brachiopods. Bony fishes** become numerous, and there are also many more kinds of **sharks.** The first **freshwater clams** appear.

Over this geologic epoch and the next one, animals lay **water-tight eggs** for the first time in Earth's history, allowing the ancestors of birds, mammals, and reptiles to reproduce on land.

Extinction: The **armored fish** become extinct.

Continental Drift

The two supercontinents Gondwana and Laurentia finally crash into each other somewhere over the equator, forming one gigantic land mass called Pangaea. This creates the Appalachian mountain range in what will be North America. Much of the land is often covered by warm shallow seas. There's less coastline than in previous periods, because the continents are all pushed together into one continent.

Weather

Tropical and humid all year 'round, even in winter.

Pennsylvanian Epoch
(also called the Upper Carboniferous)

325 million years ago–286 million years ago
325–286 MYA

Life-Forms

The Pennsylvanian Epoch sees an increase in **centipedes, spiders, giant millipedes, scorpions,** and many kinds of **cockroach. Giant horsetails, tree ferns, cycads, liverworts, club mosses,** and **scale trees** abound. There are still a few **trilobites** and **sea scorpions** in the ocean basins. There are fewer **corals, crinoids** (sea lilies), **blastoids, cryozoans** and **bryozoans** in the seas.

The Pennsylvanian is sometimes called the "Age of **Amphibians**"; some of these are huge. There are **flying insects,** including the **giant dragonflies** and **mayflies.** The **first land snails** arrive. **Early reptiles** and the **ancestors of birds** appear.

Continental Drift

The two supercontinents Gondwana and Laurentia have finally crashed into each other somewhere over the equator, forming one gigantic land mass called Pangaea. This creates the Appalachian mountain range in what will be North America. Much of the land is often covered by warm shallow seas. There's less coastline than in previous periods, because the continents are all pushed together into one continent.

Huge glaciers at the South Pole cause sea levels to drop, and the land dries up. Later, the ice melts and shallow seas flood the lands and the forests. This happens again and again, so the land goes back and forth between being flooded and drying up. Toward the end of the period, it's drying up.

Weather

Tropical and humid all year 'round, even in winter.

STUDENT HANDOUT

Permian Period

286 million years ago–245 million years ago
286–245 MYA

Life-Forms

In the Permian Period, **arthropods** such as **centipedes, spiders, giant millipedes, scorpions,** and many kinds of **cockroach** are abundant. There are still flying insects, including the **giant dragonflies.** There's a great diversity of marine life, including **foraminifers, ammonids, brachiopods, bryozoans,** and **bivalves.**

Ferns and **horsetails** decline, but **seed plants** such as **conifers** increase dramatically. There are fewer **amphibians. Terrestrial reptiles** such as **early snakes** and **lizards** flourish on land; **aquatic reptiles** begin to appear in the sea. New major **insect groups** such as **beetles, bugs,** and **cicadas** appear. There are now **freshwater bony fishes.**

Extinction: The largest mass extinction *ever* in Earth's history occurs during the Permian! As global temperatures increase, many ocean basins evaporate and disappear, and almost all **invertebrates** in the oceans go extinct—including the once-abundant **trilobites** and **sea scorpions.** On land, **diapsids** and **synapsids** (kinds of reptiles) also become extinct.

Continental Drift

During the Permian, what will be Siberia collides with the giant landmass, completing the creation of Pangaea—the largest supercontinent ever. Almost immediately, parts of the southern subcontinent Gondwana start to drift away to the north, creating new micro-continents. There are fewer glaciers in the north and south.

Weather

Over the course of the Permian, global temperatures increase dramatically. The land is humid and hot from the evaporation of the seas. Conditions are becoming more and more tropical.

Triassic Period

245 million years ago–208 million years ago
245–208 MYA

Life-Forms

In the Triassic Period, **arthropods** such as **centipedes, spiders, giant millipedes, scorpions,** and many kinds of **cockroach** are still abundant. **Turtles** appear, and **crocodilians** and **true lizards. Bony fishes, starfish,** and **sea urchins** increase, and **sharks** are plentiful.

Cycads and **ginkgoes** abound, as well as **club mosses** and **dicynodonts.** Occasionally during this period, **seed ferns** are the only plants on land. Other **seed plants** such as the true **conifers** gradually become common.

Early dinosaurs spread over the land. **Marine reptiles** such as **placodonts, nothosaurs,** and **ichthyosaurs** are common in the sea. The **therapsids,** the first mammal-like reptiles, make their first appearance. The sea level rises, allowing a recovery of life in the seas.

Continental Drift

During the Triassic, Pangaea slowly drifts to the north. The sea level rises at the beginning of the period and drops toward the end. On land the climate becomes dryer. The center of Pangaea lies across the equator, but the continent stretches all the way to the North and South Poles.

Weather

Away from the coasts, the land is very dry, with colder winters and hotter summers. The glaciers at the poles are receding.

Jurassic Period

208 million years ago–146 million years ago
208–146 MYA

Life-Forms

Large shallow seas are full of **giant marine crocodiles, sharks, rays,** and relatives of today's **squids** and **octopuses.** There are numerous **ammonites. Bony fishes** are plentiful. The great **ichthyosaurs** and long-necked **plesiosaurs** rule the seas.

On land, **beetles, snails, centipedes, spiders, millipedes, scorpions,** and many kinds of **cockroach** continue to thrive. New insects evolve, including **earwigs, flies, bees, wasps, ants,** and **caddisflies. Turtles** and **lizards** are numerous. The **earliest mammals** begin to appear. **Aerial reptiles** include **pterosaurs** such as *Rhamphorhynchus,* which has no feathers, and *Archaeopteryx,* which does. The biggest dinosaurs on land are the plant-eating **sauropods** such as *Diplodocus, Brachiosaurus,* and *Apatosaurus.* **Stegosaurs** are the plated dinosaurs. *Allosaurus* is a heavy and powerful meat eater; **coelurosaurs** are also carnivores, but light and built for speed. **Ceratosaurs,** such as *Dilophosaurus,* are very early **therapods.**

> *Note:* Although *Tyrannosaurus rex* and *Triceratops* appeared in the book and movie *Jurassic Park,* they had **not** yet evolved in the Jurassic Period!

Land plants include **conifers,** such as relatives of today's **redwood, cypress, pine,** and **yew.** There are also **cycads** and **ginkgoes, club mosses, seed plants,** and **dicynodonts.**

Continental Drift

All continents that had been joined together into the supercontinent Pangaea begin to twist apart.

Weather

The Jurassic climate is constantly warm and moist, less varied than today. During most of this period there are no icecaps covering the poles and the sea level is high. Rain begins to fall in areas that used to be deserts.

Cretaceous Period

146 million years ago–65 million years ago
146–65 MYA

Life-Forms

Until the last part of this period, the dinosaurs are still prevalent. The biggest at this time are the plant-eating **sauropods** such as *Diplodocus, Brachiosaurus,* and *Apatosaurus.* **Stegosaurs** are the plated dinosaurs. *Allosaurus* is a heavy and powerful meat eater; **coelurosaurs** are also carnivores, but light and built for speed. **Ceratosaurs,** such as *Dilophosaurus,* are very early **therapods.** *Tyrannosaurus rex* and *Triceratops* appear during the Cretaceous, as well as **duck-billed** and **horned dinosaurs.** **Pterosaurs** soar in the skies, including one of the biggest flying dinosaurs ever, *Quetzalcoatlus.*

Amphibians, lizards, crocodilians, and **snakes** continue to increase. **Ants, butterflies, aphids, grasshoppers, gall wasps, termites,** and a kind of **bee** appear for the first time. This period sees the evolution of the first **modern birds,** including **gulls** and **wading birds.** The first **modern mammals** also appear.

In the sea, **diatoms, marine reptiles** such as **ichthyosaurs** and **plesiosaurs,** and **giant marine crocodiles, sharks,** and **rays** are abundant. There are also **ammonites** and **nautiluses,** and relatives of today's **squids** and **octopuses. Bony fishes** are everywhere.

The first **flowering plants** appear, including **magnolias** and **palms.** They establish themselves in the damper regions. There are still **seed ferns, cycads,** and **conifers** in great forests.

Extinction: The most famous (but not the largest) extinction on Earth occurs at the end of the Cretaceous. **Dinosaurs** die off on land. **Pterosaurs** become extinct, but birds remain. In the sea, many **foraminifers** and **ammonites** die off, along with many **marine reptiles—ichthyosaurs, plesiosaurs,** and **mosasaurs.**

Continental Drift

The continents continue to move apart, and are separated by ocean basins. Africa separates from South America. Europe separates from North America. Africa, India, and Australia move north. The Rocky and Andes mountain ranges are forming. There is evidence of a huge asteroid hitting Earth at the end of the Cretaceous.

Weather

Seasons begin to grow more distinct as the global climate cools down.

Tertiary Period

65 million years ago–1 million, 800,000 years ago
65–1.8 MYA

Life-Forms

The first **flowering plants** have appeared, including **magnolias** and **palms.** There are still **seed ferns, cycads,** and **conifers** in great forests, and broad-leafed **deciduous trees.** More open habitats have evolved—**grasslands, tundra,** and **desert.**

Amphibians, lizards, crocodilians, and **snakes** continue to increase. **Ants, butterflies, aphids, grasshoppers, gall wasps, termites** and other insects are flourishing. **Modern mammals** have taken hold, including **marsupials, insectivores,** and **rodent-like animals.** With more open habitats, the first hoofed running and grazing animals, like **early deer, antelopes, pronghorns, gazelles,** and **horses,** evolve. More **large hunting mammals** (including the **saber-toothed cat**) and **hunting birds (raptors)** take advantage of these new populations. The first **elephant-like animals** appear, as well as **dogs, cats, pigs, giraffes,** and **weasels. Early bats** appear. There are **mastodons** on the land. The first **primates** appear.

In the seas, the first **completely marine mammals** evolve. These are similar to modern **whales, porpoises, dolphins,** and **manatees. Bony fishes, sharks,** and **rays** are abundant, including the giant shark, *Carcharodon megalodon.* **Seals** and **walruses** appear. There are also **nautiluses,** and relatives of today's **squids** and **octopuses. Kelp forests** evolve in the sea.

Extinction: The most famous (but not the largest) extinction happened at the end of the Cretaceous, just before this Tertiary Period. **Dinosaurs** died off on land (but not birds), and in the seas many **foraminifers** and **ammonites** died. During the Tertiary, many species near the North and South Poles became extinct.

Continental Drift

A land bridge forms between North and South America, and animals can now travel between them. India crashes into Asia, forming the world's tallest mountains, the Himalaya. Australia separates from Antarctica and moves north. Africa meets up with Europe, creating the Alps and the Pyrenees. By the end of this period, the continents look much as they do in modern times.

Weather

Temperatures first warm, then cool, then warm again. Away from the coasts, the lands gets dryer. The North and South Poles build up a lot of ice, as glaciers alternately cover much of the planet, melt, and freeze again.

Quaternary Period

1 million, 800,000 years ago–present day
1.8 MYA–present day

Life-Forms

All plants and animals living in the present already exist in the Quaternary. The **woolly mammoth** appears, as well as **giant ground sloths, giant beavers,** and **long-horned bison.** Great **teratorn birds,** with 25-foot wingspans, fly above Earth. **Flowering plants, birds,** and medium and small **mammals** continue to increase and expand their ranges. **Early humans** appear at the very, very end of the period. As land bridges form between major land masses, humans and other animals migrate between continents.

Extinction: Over the course of the Quaternary Period there's a gradual decline in the last **giant mammals.** About half-way through, the **mammoths, saber-toothed cats, giant ground sloths, giant beavers, long-horned bison,** and **teratorn birds** become extinct. "Background extinctions" of many species on land, in the air, and in the sea continue to the present day.

Continental Drift

In the Quaternary, the continents have formed as they appear in modern times. During the ice ages, so much of Earth's water freezes that the sea levels drop. With lower seas, land bridges appear between North America and Asia, and in other parts of the world.

There is evidence that the continents seem to be moving toward one another again, and may join to form a new supercontinent sometime in the future, this time near the North Pole.

Weather

The weather has alternated between very cold during the ice ages, when glaciers covered a lot of the land, and warm, when the glaciers melted. In the present day, the temperatures appear to be going up...which could have a significant impact on life on Earth.

Explaining Major Evolutionary Change

1. Look for **EVIDENCE** that the change you're writing about actually occurred, referring to the "Class Time Line" and the "Life through Time" wall chart.

2. Look at the **MOVEMENT OF CONTINENTS** before and during the change. What clues does this give you?

3. Look for **CHANGES IN OTHER ORGANISMS** before and during the change. What clues does this give you?

4. Look for evidence of **CATASTROPHIC EVENTS** before or at the time of the change. What clues does this give you?

5. Brainstorm the kind of **EVIDENCE** needed **to support your explanations.**

6. **Write up your ideas** in the form of a report or presentation. If you don't have the necessary evidence for your ideas, explain what you would look for as a scientist investigating this question.

"If the Eiffel tower were now representing the world's age, the skin of paint on the pinnacle-knob at its summit would represent man's share of that age; and anybody would perceive that that skin was what the tower was built for. I reckon they would. I dunno."   —*Mark Twain*

The history of life on Earth is one of the great detective stories in the Universe. Over the course of 3.5 billion years so far, life has emerged, adapted, flourished, been extinguished, evolved differently, declined, flourished, and evolved differently again in a never-ending cycle of adaptation. And although we often treat the planet as if we've been on the scene (and controlled its disposition) forever, *Homo sapiens* appeared only in the *very, very* recent past, mere bats of an eyelash ago, adding to the variety and adaptive marvels of life on Earth today.

The "deep time" of Earth history (a phrase coined by author John McPhee) tells its own detective story. Multicellular life exploded on the scene, fish evolved jaws, dinosaurs squished puny mammals between their toes, and a tiny, insignificant group of primates (our ancestors) left their forest homes for life on the open grasslands of what is now Africa. These events are real; they happened. However, unlike the case of human history, no one was around to document or record these events. And unlike the recorded blip of human history (perhaps 6,000 years to date), the 3.5 billion (3,500 million) years during which life has been around on Earth is an incomprehensibly long time. The evidence available—the fossil record, and anatomical similarities between living species—allows scientific historians (paleontologists) to sift through the record to interpret and make sense of how life evolved over time.

A Full-Bodied Science

"In science, a theory is not a guess or an approximation but an extensive explanation developed from well-documented, reproducible sets of experimentally derived data from repeated observations of natural processes."
—*National Association of Biology Teachers (NABT)* Statement on Teaching Evolution, *1995*

Modern evolutionary theory involves the skills of many disciplines: geology, biology, paleontology, genetics, ecology, biochemistry, developmental biology, molecular biology, anthropology, even physics. Corroborative findings among these disciplines, as well as extensive observation, experimentation, and creative reflection, support evolution as one of the strongest and most useful scientific theories of our time. With so much supportive evidence and data available, it can seem daunting for a classroom teacher to know where to begin.

This background material is intended to make sense of significant and reoccurring key concepts presented in *Life through Time*. Rather than being divided into strict, session-specific points, the background is organized into major evolutionary themes that may present themselves during *any* of the sessions. Some topics expand concepts that are just touched upon in the activities. Others address misconceptions often associated with the subject matter. We've attempted to present the most current understandings of each topic without getting bogged down by too many details; additional resources are provided for further inquiry. You may wish to read "Important Evolutionary Themes" on page 289 before continuing. You'll see many of those points revisited here, with more ample explanations.

Evolution Is Unpredictable

Looking at the world around us, at the dizzying diversity of organisms that live (or have ever lived), it's tempting to take away the impression that given enough time, evolution can build just about anything. After all, the planet's seen everything from jellyfish to dinosaurs, from mosses to peacocks. There are close to 19,000 different species of orchid, over 20,000 different species of fish, and well over three million beetle species identified so far.

But evolution is not an inevitable march towards humans, pandas, maple trees, and elephants. It merely appears this way in retrospect.

If we traveled back in time and stood on the shores of a Precambrian sea, could we predict that the bacterial and algal stromatolite mounds would transform the atmosphere into an oxygen-rich world? Could we know which Cambrian multi-celled life-forms would survive and which would disappear? Could we identify the potential for diversity in our early chordate ancestors? And if we looked out over a Cretaceous

*The National Science Education Standards recognizes that conceptual schemes such as evolution "unify science disciplines and provide students with powerful ideas to help them understand the natural world," and recommends evolution as one such scheme. In addition, the Benchmarks for Science Literacy from the American Association for the Advancement of Science's Project 2061 and NSTA's Scope, Sequence, and Coordination Project, as well as other national calls for science reform, all name evolution as a unifying concept because of its importance across the discipline of science. Scientific disciplines with a historical component, such as astronomy, geology, biology, and anthropology, cannot be taught with integrity if evolution is not emphasized.
— from the National Science Teachers Association (NSTA) Position Statement on teaching evolution, 1997*

*Biological evolution refers to the scientific theory that living things share ancestors from which they have diverged: Darwin called it "descent with modification." There is abundant and consistent evidence from astronomy, physics, biochemistry, geochronology, geology, biology, anthropology, and other sciences that evolution has taken place.
—NSTA, "The Nature of Science and Scientific Theories"*

landscape, could we possibly surmise that dinosaurs would soon become extinct not because of predation, competition, or disease, but as a result of an asteroid impact from space? When seen through this lens, the history of life on Earth appears less predictable and determined—and more a series of contingent and fortuitous events. **The history of life makes sense only in retrospect; our world today is just one possible outcome out of countless others that could have resulted.**

Why Evolution Happens

Evolution is the "net process" of change occurring in a species over time. While this process is evident from studying the fossil record and observing similarities and differences among living species, the actual mechanisms of evolution are less clear. The classic Darwinian model involves the mechanism of **natural selection.** Darwin observed that resources are *limited* in most ecosystems, and that there exists a *struggle* for survival among living things. He also observed that certain *heritable* traits (adaptations passed down in the genes) could provide selective advantage in this struggle. (A certain adaptation worked really well for great, great, great, great, great-grandfather mole, so the trait was passed along through the generations.) And these heritable traits can *vary* from offspring to offspring. Darwin's triumph was piecing his observations together into an elegant mechanism for change: individuals whose traits (adaptations) are best suited for survival will out-compete (be selected over) other individuals, and will successfully transmit these traits along to their offspring. Over time (many generations), new species may arise.

Building on What Comes Before

The evolutionary process involves the interplay of organism and environment. The environment exerts pressures, and evolution shapes and builds on what already exists. An organism can't suddenly evolve feathered wings unless precursors exist for this structure (bones, muscle, nerve tissue, feathers, etc.). The more variety there is at a given time, the more variety there can be in the future. Organisms in the relatively short recent geologic past (the past 65 million years or so) are dramatically more diverse than they were in the long early millennia of life on Earth. Again, while the tendency appears to be an increase in complexity and diversity over time, the end result is unpredictable.

Natural selection, the primary mechanism for evolutionary changes, can be demonstrated with numerous, convincing examples, both extant and extinct. —NABT

Deep Time

In *Life through Time*, the life changes are continuously connected with geologic history and time. Geologic time is immense. The Earth is approximately 4.5 *billion* years old. How do we wrap our brains around this enormous a number? Many authors have tried to illustrate geologic time by compressing the 4.5 billion years into smaller, more human scales—metaphorically equating it to a year (365 days), for instance, or an hour (60 minutes), or even a yard (36 inches). Even with metaphor, though, the depth of geologic time is difficult to fathom. Whether years, months, seconds, or inches are used, Earth time still seems elusive.

The Story in the Rock

No rocks on Earth can be dated back 4.5 billion years, because early in its history the Earth's surface was scalding hot, and the original rocks melted and were later re-formed into other rocks. So how do we know Earth is older than that? Our Moon, which was formed at the same time as Earth but didn't heat up as our planet did, has rocks that can be dated to approximately 4.5 billion years ago. Other debris in our solar system has also been dated at 4.5 billion years old. This is when scientists think that Earth, along with the rest of the solar system, formed.

The geologic time scale represents the best method for placing the planet's historical events in perspective. Since no single location on Earth displays a continuous sequence of rock from Precambrian to the present, the geologic time scale consists of rock strata figuratively assembled from different locations around the world. But how was the sequence assembled? The rocks were "relatively" dated in the 1800s, using "index fossils" (commonly found, widely distributed fossils that are limited in time span) in the rock. Absolute dating (and confirmation of the original relative dates) became possible at the turn of the 20th century, shortly after the discovery of radioactivity. By measuring the quantities of radioactive elements (and the elements into which they decay) in rocks, scientists can determine how much time has elapsed since the rock initially formed.

An interesting way to think about geologic time is to compress it into a more familiar measure. If Earth's history is "capsulized" into one year starting January 1st, the most simple plants don't arrive until November, and recorded human history occurs only in about the last five minutes before midnight on December 31! On the Web, at www.athro.com/geo/timecalc.html, students can use the "geological time scale metaphor calculator" to pick different measures (such as meters, days, minutes), then automatically calculate where major events in Earth history would have taken place.

Geologic Time Scale

Time	Era	Period	Epoch
Precambrian	Archean		
	Proterozoic		
Phanerozoic	Paleozoic	Cambrian	
		Ordovician	
		Silurian	
		Devonian	
		Carboniferous	
		Permian	
	Mesozoic	Triassic	
		Jurassic	
		Cretaceous	
	Cenozoic	Tertiary	Paleocene
			Eocene
			Oligocene
			Miocene
			Pliocene
		Quaternary	Pleistocene
			Holocene

GEOLOGIC TIME

eras	periods and systems
Cenozoic	Quaternary
	Tertiary
Mesozoic	Cretaceous
	Jurassic
	Triassic
Paleozoic	Permian
	Upper Carboniferous ("Coal Measures") Carboniferous Lower Carboniferous ("Mountain Limestone")
	Devonian ("Old Red Sandstone")
	Silurian
	Ordovician
	Cambrian
Precambrian	

A classic textbook geologic time scale. The Precambrian should represent 90 percent of the whole time line!

How the Story Is Divided

The divisions within the standard geologic time scale represent significant "moments" in Earth's history. Divisions in the scale move from large blocks of time to smaller ones. The largest divisions, **eons** and **eras,** represent enormous stretches of time (billions to hundreds of millions of years). **Periods** are shorter, and typically signal the extinction, rapid diversification, or first appearances of organisms. (Period names often refer to the geographic location where the rock types were first discovered; Cambrian, Ordovician, and Silurian refer to the regions of Wales where the first rocks of each period were identified.) The smallest divisions of the scale, **epochs** and **ages**, allow geologists to isolate extremely precise moments in geologic time (upper, middle, and lower periods) or even narrow the time down to a few thousand years.

Unfortunately, in most textbook geologic time scales the largest block of Earth's history, the Precambrian Time, is depicted as the smallest section at the very bottom of the scale. This is extremely misleading; the Precambrian represents 90 percent of the entire history of Earth! In *Life through Time,* the class time line you construct throughout the unit depicts more realistic proportions…but even at that, it's virtually impossible to convey the vast expanse of the Precambrian. A correctly proportioned geologic time scale would either include several pages dedicated to the Precambrian, or force the compression of 3 billion, 956 million years of multicellular life into a miniscule space at the very top of the scale.

How the Continents "Fit"

Stare at a globe or world map long enough and you begin to notice the peculiar jigsaw puzzle–like fit of the continents. Looking at the world from the South Pole, it almost looks as if all the continents of the Southern Hemisphere are exploding away from Antarctica. This curious fit of the continents became apparent to Dutch mapmaker Abraham Ortelius in 1596, and again to many observers of the late 19th and early 20th centuries after adequate world maps had been created. But while it was one thing to notice the potential fit of the continents, it was an entirely different task to try to explain it.

Alfred Wegener, a brilliant, multidisciplinary scientist of the time, set out to do it. He devised the mechanism called **continental drift.** He argued that all present-day continents were pieces of what had been a phenomenal "supercontinent," **Pangaea** (Greek for "all land"), which began breaking up 200 million years ago. Matching rock types and fossils from different continents, Wegener assembled considerable evidence that supported his explanation, and published *The Origin of Continents and Oceans* in 1915. What he lacked, however, was an explanation for the actual *movement* of the continents. As a result, his idea, though compelling, lay dormant for almost 50 years.

Shifting Plates

It took new research and technologies from the mid-1950s through the 1970s for scientists to vindicate Wegener's original notion of moving continents. They did it with a working mechanism called **plate tectonics.** Deep-sea exploration revealed areas of ocean floor where new molten rock was pushing apart the pieces of Earth's surface—the mid-ocean ridge. As the tectonic plates that make up the Earth's crust shifted, some collided with each other, forming mountain systems and causing volcanic eruptions and earthquakes across the planet. Some plates were forced beneath other plates in a process called subduction. Plate movements even caused continents to tear apart.

The GEMS guide Plate Tectonics, *for grades 6–8, beautifully explains and engages students in the concepts of plate movement, continental drift, and geologic time.*

Over millions of years, plates can migrate on the Earth's surface, altering the positions of the continents. This has a direct influence on the possibilities for the history of life, as well. The position of the continents dictates the local **climate**—warmer at the equator, colder farther away. That, in turn, influences how species evolve; species located near the equator will face radically different selective pressures from those in higher latitudes. The movement of plates can also affect **ocean currents,** a major factor in global climate regulation. Continents can block or divert ocean currents, causing drought in some areas and monsoon-like moisture in others. Enormous changes occur when **separate species** are brought together as continents collide. Whole **habitats** can also be created or destroyed, such as when vast, warm, shallow ocean basins covered much of North America, and again when they were annihilated as the continents shifted again.

Ocean Currents, *the GEMS marine-science teacher's guide for grades 5–8, explains the critical role of marine circulation on Earth.*

Fossils

The word "fossil" has been defined in many different ways. It can simply be defined as any evidence of prehistoric life. This definition has the benefit of being very simple and also very broad. Since many parts of an animal or plant may not be represented in the actual fossilized remains, paleontologists must reconstruct an organism from all *available* evidence. Such reconstructions obviously include all the direct evidence of prehistoric life—dinosaur bones, petrified wood, amber, teeth, and shells—but also the indirect evidence, such as footprints, traces, or holes in mudstone, and coprolites (fossilized feces). And fossils can include much more than just bones, teeth, wood, and feces. Individual cells, pollen, and spores can actually be identified and isolated from the surrounding sediments that buried them. Whether direct or indirect, a fossil is evidence of a once-living organism on Earth.

What Makes a Fossil?

From the enormous collections of fossils in museums around the world one might think that fossilization is a common occurrence. But the process itself is incredibly rare. Think about your last walk through a forest. Can you recall stumbling over the carcass of a dead deer? a bear? a mouse? They certainly are out there, but resources such as animal carcasses are recycled very quickly, and the likelihood of finding one is slim. Remains that aren't eaten are likely scattered, and these fragments decompose quickly under the influence of oxygen and weather. In an organism with no bones, soft body parts are eaten or quickly decompose, and any hard parts are eaten, crushed, scattered, or dissolved. All in all, the likelihood of specimens becoming fossilized is very small.

For fossilization to have a chance to occur, animal and plant remains must be somehow insulated from the environmental conditions that normally accelerate decomposition. Most fossilization involves animal or plant remains being quickly blanketed by some sediment (sand, mud, ash) that shields the organic material from decay. Over time, minerals from the sediments take the place of organic material in the buried animal or plant. The rapid burial/submergence scenario necessary for fossilization is very rare, and as a result, few organisms become fossilized.

Punctuated Equilibrium: Evolution in Fits and Starts

The classic model of natural selection is based on Darwin's premise that change occurs gradually over extremely long stretches of geologic time.

Darwin recognized that the fossil record didn't support this gradual, continuous scenario, and claimed that gaps in the sequences of fossil-bearing rock meant that important transitional forms were missing. These gaps were interpreted as "missing links" in a slow, continuous process of change over time.

For more than a century after Darwin proposed his theory, paleontologists struggled with a so-called incomplete fossil record. Even more confusingly, the fossil record itself seemed to contradict Darwin's premise that species underwent slow change throughout their lineage. On the contrary, fossil species appeared unchanged over extremely long stretches of geologic time and then underwent rapid episodes of anatomical change. In 1972, invertebrate paleontologists Niles Eldredge and Stephen Jay Gould caused a commotion in the evolutionary community by challenging classic Darwinian gradualism.

Eldredge and Gould reinterpreted the tempo and distribution of evolutionary change as a process they called **punctuated equilibrium** (or "Punk Eek," for short). They observed that the fossil record indicates long periods of stability (equilibrium) interrupted by bursts of transformation ("punctuations" in the overall equilibrium). Remember, fossilization is an iffy process. And episodes of transformation occur in the geologic blink of an eye (perhaps 5,000 years—which barely represents a single layer of rock in the fossil record). So the fossil record can *appear* "incomplete." Yet extremely rare transitional fossils do exist, such as the famous *Archaeopteryx,* which has both reptilian and birdlike structures, and the stunning whale fossils from Pakistan showing the intermediate, legged ancestors of modern whales. Punctuated equilibrium takes the fossil record at face value and accounts for the apparent "gappiness" in the record: if there's evidence of transitional organisms, transition occurred. More than thirty years after its publication, punctuated equilibrium remains a powerful and influential argument in modern evolutionary theory.

The model of punctuated equilibrium provides another account of the tempo of speciation in the fossil record of many lineages: it does not refute or overturn evolutionary theory, but instead adds to its scientific richness. —NABT

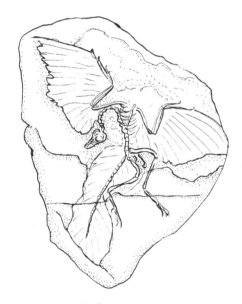

Archaeopteryx

Evolution Means Change, not "Progress"

In his life's work of 1859, *The Origin of Species,* Charles Darwin laid out a carefully crafted argument for a mechanism of evolution. It stands today as the foundation of modern evolutionary theory. Yet, in his masterpiece, Darwin struggled painfully over whether to use the word "evolution" at all. In fact, he uses the word only once—it's the very last

"As mankind has advanced, so have his tools. And thus the Verizon Yellow Pages ...."

word of the book. Why would Darwin, so convinced of his conclusions, waver at using the word?

In Darwin's Victorian England, "evolution" meant an unrolling (almost literally) of predetermined events—an unwavering movement toward progress and improvement. Darwin refuted this perception; neither the fossil record nor his observations showed evidence that life *improved* through time, only that it *changed* through time. The notion of natural selection was **not** a notion of "progress."

Darwin was a radical and progressive thinker, but he eventually capitulated to societal pressure. In *The Origin of Species,* he not only used the word evolution to represent the going definition, but even included a reference to life's progress over time. As Darwin feared, reviewers of *Origin* commented on the "progress" of biological evolution, and even used his "natural selection" model to rationalize social and class structure and injustice in Victorian society. Even today, we're surrounded by evidence that many still equate the idea of "evolution" with progress— even to imply an improvement over previous versions of a product!

What Makes for Success?

Single-celled life is very simple. A salamander or a petunia is extremely complex. But each is very well adapted to its own particular niche. Reptiles, horses, and humans are not more "advanced" than single-celled organisms; they merely have a greater complexity of cellular arrangement, functions, and behaviors. **Complexity** is one measure of the relative success of a species, and one barometer for success in the history of life.

Another factor is **diversity.** Humans are a single species, the sole surviving members of a previously more robust (but still not very diverse) group of primates called hominids. On the diversity scale, that's not too impressive. On the other hand, 80 percent of all multi-celled life is found within a single phylum: Arthropoda (insects, crabs, barnacles, lice, spiders, and their kin). Clearly, humans are minor characters in their world.

Another criterion for success is **longevity**—not how long an individual lives, but how long a particular lineage remains a player in the game of life. The single-celled organism is more successful in terms of sheer numbers and how long it's been around than any other organism on

Earth. (There are more bacteria in your mouth right now than there have ever been humans alive on the planet!) In that sense, the single-celled organism could be called life's greatest success story.

"Recent" arrivals such as arthropods, sharks, and flowering plants have been around for hundreds of millions of years and are still going strong. Even a lineage that fizzled out, the dinosaurs, had an uninterrupted 150 million–year run (and if you consider birds' apparent relatedness to dinosaurs, you can argue that the dinosaurs are *still* strutting their stuff). Our human species, having arrived on the scene a geologic nanosecond ago, has a track record of only one million years. Are humans a success story? Check back in 100 million years.

Size Matters

Over two-thirds of life's history on the planet is minute and single-celled. Even with the astonishing diversity of life on Earth today, the most common organism on the planet is the single-celled bacterium; being single-celled is obviously a successful strategy for making a living. It starts to make sense that multi-celled organisms took so long to make an appearance in the history of life.

There certainly are some big advantages to being large. The most important? Fewer predators. If you're larger than your predator, you force the predator to evolve to be bigger too. (Evolutionary biologist Geerat Vermeij called this process **"escalation."**) To successfully prey upon you, predators need to keep pace in size over generations, or evolve other adaptive feeding strategies over generations…or else go extinct.

What Keeps Size in Check?

Single-celled life is tiny—so tiny that tools such as microscopes are needed to see individual organisms. Furthermore, they'll always be tiny. We'll never see single-celled bacteria or protozoans the size of beach balls, because the upper range of their size is set by some inflexible foundations of math and physics.

Cells (like bacteria or protozoa) have a greater surface area than they do volume. There's more body surface than internal stuff. Their insides are never too far away from the outside world. Food, water, and oxygen can easily travel in and wastes can be quickly eliminated. As a cell increases in size, however, its volume quickly surpasses its surface area. Suddenly, there's more volume (stuff inside) than there is body surface to handle it. Getting material in and out of the cell gets difficult, and the cell starves or becomes overrun with cellular wastes.

So, to get bigger, you need an **increase in cells**. But that's not the only contributor to success. Sponges are composed of thousands to millions of cells, but every cell is doing essentially the same thing. More complex organisms have cells organized into **divisions of labor.** Some procure food, some store energy, some eliminate waste, some secrete calcium carbonate shells for protection. Through this **complexity** and division of labor, large organisms as diverse as cactus, orangutans, oysters, and elephants can succeed even though the individual cells they're made up of are tiny.

But getting large is also costly. Even with division of labor among cells, there are limits to and problems with getting big. Ants, spiders, and preying mantids—contrary to numerous 1950s horror movies—can never exist much larger than they are now. Again, the surface area to volume ratio limits the size of these arthropods. Lacking an internal skeleton for support and muscle attachment, big bug exoskeletons would either be incapable of movement or would shatter beneath the creatures' own body weight. Their method of breathing (small holes along their sides to extract oxygen from the air) couldn't supply oxygen to a much bigger bug. So even if a monstrous ant could move, it would suffocate long before taking its first step.

Gravity Affects Form

The force of gravity on Earth has a direct effect on biological architecture. An increase in size requires a compensatory change in form. Ants, with their large surface area and small volume, can race along on spindly, jointed legs set along the sides of their bodies. Elephant legs, on the other hand, are short and column-like to support the massive, lumbering volume directly above. If an ant could possibly overcome all other obstacles and actually attain the size of an elephant, it would no longer look like an everyday ant. Its legs would need to shorten and thicken to compensate for its mass; its weight would be centered directly over its legs; and its movement would be slower. In short, it would start to look and behave more like an elephant than an ant.

Similar to...Related to...Descended From

Common structures (also known as **homologous** structures) are important evidence of evolution. Homologous structures share the same developmental and evolutionary origin; for example, human hand bones correspond to bones in bat wings. Similarities can be observed in external features, but are even more apparent with internal organs, chemical processes, and embryonic development. In *Life through Time* students see a vast array of diverse organisms, but the focus is on evidence of **relatedness**—the similarity of such characteristics as digestive tracts and reproductive strategies.

Surveying the diversity of life on Earth, certain themes seem to crop up time and again. For example, dolphins, ichthyosaurs (extinct marine reptiles), and sharks all display similar hydrodynamic, streamlined bodies, yet have very different vertebrate origins (they're mammals, reptiles, and fish, respectively). Flight has evolved independently over and over in widely divergent groups of organisms. Insects, birds, pterosaurs (reptiles), and bats (mammals) all fly, but they employ radically different anatomy.

Similar adaptations can develop more than once in the history of life, by remotely related lineages. This process is called **convergence.** For the aquatic species listed above, their common body shape came of the similar environmental pressures of making a living in an aquatic habitat. Similarly, overcoming the force of gravity and the overall demands of flight dictate a similar solution across different species.

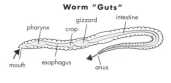

Worm "Guts"

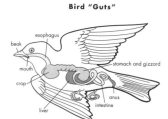

Bird "Guts"

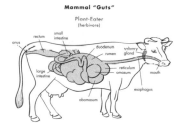

Mammal "Guts"
Plant-Eater (herbivore)

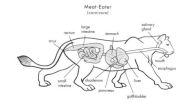

Mammal "Guts"
Meat-Eater (carnivore)

Life-Form Explosions...

Imagine the vast, empty seas of the Precambrian. For 3.2 billion years, this marine world contained only single-celled life. Nothing more. Then, in a snap (perhaps 40 million years), every basic multi-celled anatomical design exploded on the scene. The so-called Cambrian Explosion marked the beginning of almost every major group of modern animals. (It also gave rise to anatomical forms so strange that they continue to defy classification.) Not a single new anatomical design (with the single exception of bryozoans in the Ordovician) has appeared in the 544 million years since the Cambrian.

How could multi-celled organisms suddenly appear from a sea of single-celled life? As paleontologist Stephen Jay Gould has put it,

When compared with earlier periods, the Cambrian Explosion evident in the fossil record reflects at least three phenomena: the evolution of animals with readily fossilized hard body parts; Cambrian environment (sedimentary rock) more conducive to preserving fossils; and the evolution from pre-Cambrian forms of an increased diversity of body patterns in animals.
—NABT

"Geological explosions have awfully long fuses." As discussed earlier, fossilization is a relatively rare event, and fossilization of soft anatomy is almost unheard of. Yet a few extremely rare fossils from the Vendian Epoch, almost 100 million years before the Cambrian Explosion, show very clear evidence of multi-celled life. Some of these forms have been identified as early jellyfish and flatworms. Others are absolutely bizarre and relate to no known modern group of animals. What these fossils record is the earliest preserved evidence of life's attempt at multicellularity.

...and Extinctions

At the other extreme from sudden, explosive appearance in the fossil record is the abrupt, sometimes cataclysmic disappearance of species. While "background extinctions" occur constantly due to habitat loss and over-harvesting of species, the history of life is also punctuated by several major disruptions known as "mass extinctions"—the elimination of major plant and animal groups in a geologic instant (perhaps over the course of a million years; frequently far more suddenly). Paleontologists have identified five mass extinction events, marking the ends of the Ordovician, Devonian, Permian, Triassic, and Cretaceous Periods (the last being the so-called "KT" Event). Though each of these mass extinction events had significant implications for the subsequent history of life on Earth, *Life through Time* focuses most directly on the periods surrounding the Permian and KT extinctions.

Best estimates indicate that approximately 225 million years ago the Permian extinction wiped out *96 percent of all marine species,* and almost derailed even the mighty arthropod supremacy. Explanations for the Permian extinction have ranged from lethal radiation from a nearby supernova (the explosion of a dying star) to more recent evidence of impact by an asteroid. Tellingly, the Permian Event coincides with the largest volcanic activity in Earth's history.

The KT Event happened approximately 65 million years ago, at the end of the Cretaceous and beginning of the Tertiary Periods. It marked the end of the dinosaurs and the beginning of mammal diversification. The bulk of evidence indicates that an extraterrestrial body such as an asteroid collided with Earth near what is now Mexico's Yucatan Peninsula. Debris and dust from the impact traveled high enough in the atmosphere to block sunlight across the planet. The cooling effect and collapse of ecosystems all over the planet marked the end of the line for dinosaurs…but just the beginning for a small group of mammals.

These mass extinctions imply that the history, tempo, and direction of life are unpredictable. Gould has observed that there's a lot of quirkiness to life's history: "If dinosaurs had not died in this event, they would probably still dominate the domain of large-bodied vertebrates, and mammals would still be small creatures in the interstices of [the dinosaurs'] world."

Plant and "Plantlike"

In *Life through Time*, the first photosynthesizing life on Earth is referred to as "early plants." This needs clarification. "Early plants" were photosynthesizers, but not "plants" in the strict sense of the word. They were more closely related to bacteria than to true plants. So what does it take to be a plant? No one would disagree that mosses, redwoods, petunias, and palm trees are plants. They're all organisms associated with the condition of being a plant: they photosynthesize, are multi-celled, have cell walls composed of cellulose, and store their energy in the form of true starch. But many other photosynthesizing organisms don't fit these criteria at all.

Organisms that employ solar energy and chlorophyll to produce their food are often referred to as plants, but photosynthesis isn't all there is to plant-hood. Unless those organisms have true roots, stems, leaves, and vascular tissue to transport nutrients and water, it's more accurate to refer to them as "plantlike." Seaweeds (and other forms of algae) are often casually labeled plants as well, though again, it's more accurate to refer to them as "plantlike."

Are Birds Related to Dinosaurs?

One of the most exciting moments in science is witnessing a significant shift in the scientific community's understanding or acceptance of a theory. It doesn't happen often (or easily). But when a "paradigm shift" does occur, it's big news. The last half of the 20th century saw just such a shift regarding the classification of dinosaurs.

Dinosaurs were originally classified as reptiles. (Enormous ones.) But as **ectotherms,** reptiles depend on their environment to maintain body

The models and the subsequent outcomes of a scientific theory are not decided in advance, but can be, and often are, modified and improved as new empirical evidence is uncovered. Thus, science is a constantly self-correcting endeavor to understand nature and natural phenomena. —NABT

temperature sufficient to support their metabolism. Some big reptiles (crocodiles, Komodo Dragons, and pythons) attain jumbo proportions—but at a cost. They're restricted to tropical climates and have to bask each day to raise their body temperature. A sudden drop in temperature renders them inactive. A sustained cold snap can spell death. While some internal heat could have been generated from dinosaur size alone, that possibility was never thought to fully account for a reptile attaining the size of a Brontosaurus.

In the 1970s, Robert Bakker and other paleontologists transformed the image of dinosaurs from slow-moving ectotherms to, in at least some cases, **homeothermic** ("warm-blooded") giants well equipped to roam the Earth for 150 million years. Bakker revised the classification of dinosaurs, repositioning them as the ancestors of birds. This places dinosaurs in a specific subgroup of the archosaurs, a group (clade) that includes crocodiles and birds.

What Were the Feathered Dinosaurs?

In one of the most electrifying discoveries of the last century, fossils unearthed in northeastern China revolutionized our understanding of the origin of feathers and flight in dinosaurs. Exquisitely preserved skeletons of four new theropod ("beast-footed") dinosaurs—*Sinosauropteryx, Protarchaeopteryx, Caudipteryx* and *Confuciusornis*—bear remains of feathers, although only one of them (*Confuciusornis*) could actually fly. These early Cretaceous fossils reveal three different stages in the evolution of the shoulders and forelimbs of theropod dinosaurs and provide crucial evidence for how grasping arms transformed into flying wings. And long before any feathered fossils were found, Bakker hypothesized that some dinosaurs were equipped with insulating feathers.

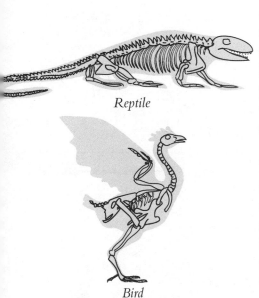

Reptile

Bird

The debate on these issues continues, sometimes hotly, but new discoveries and research lend strong support to these views. Many paleontologists, taxonomists, and other scientists who study this question feel that there is strong, compelling evidence that birds evolved from small, terrestrial theropod dinosaurs in the group Maniraptora. Maniraptorans have many structures in common with birds, including modified elements in the wrist, which makes the flight stroke possible and which probably evolved from a grasping function (Maniraptora means "seizing hands"). They also share a fused clavicle ("collar bone") and sternum ("breast bone"), long "arms," a unique pelvic structure, and other common adaptations.

Archaeopteryx

Archaeopteryx is considered by many to be the first bird, a transitional form between birds and reptiles dating back approximately 150 million years. It's thought to bear even more resemblance to its ancestors, the Maniraptora, than to modern birds. The substantial evidence seems to indicate that birds descended from these early reptiles—or at the very least, from a common ancestor.

What Defines an Adaptation?

Adaptations are features or behaviors that can improve an organism's chance for survival. The colonization of new habitats (such as dry lands), was only made possible by new adaptations ("watertight" skin, a hard-shelled egg, a water-conserving large intestine, etc.). These adaptations occur over many, many generations in a lineage; **an individual organism cannot adapt its features within its lifetime.**

Tempting as it may be to view every feature of an organism as an adaptation, it's likely that many features had no adaptive significance whatsoever—or that the original adaptation was important when the organism first evolved but is no longer valuable (as with the human appendix—as far as we know). Structures that serve one function may be "co-opted" for some novel, unrelated function. Natural selection can further modify the original structure to better serve the second function. So, wings didn't evolve initially for flight, but flight is now the latest use in a continuing series of adaptative experiments, from climbing to parachuting to gliding. Likewise, the eye has its precursors in early light/dark sensors of flatworms. And some of the complex biochemical pathways in humans bear little or no resemblance to the chemical reactions from which they evolved.

> *Adaptations do not always provide an obvious selective advantage. Furthermore, there is no indication that adaptations— molecular to organismal—must be perfect: adaptations providing a selective advantage must simply be good enough for survival and increased reproductive fitness. —NABT*

Organizing the Sequence of Life

Humans (like all primates) are incredibly visual creatures. The bulk of what we perceive about the world comes to us through our eyes. Because of this, visual input (charts, graphs, tables) can pack a wallop of information. Not surprisingly, scientists have attempted to convey the astonishing diversity and connectedness of life on Earth through pictures or icons. Many of these attempts ended up conveying cultural stereotypes, biases, and misrepresentations. To Aristotle and his followers, for

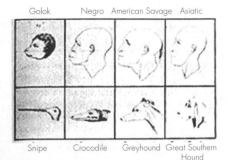

Golok   Negro   American Savage   Asiatic

Snipe   Crocodile   Greyhound   Great Southern Hound

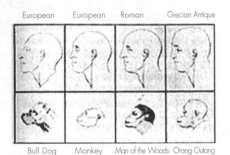

European   European   Roman   Grecian Antique

Bull Dog   Monkey   Man of the Woods   Orang Outang

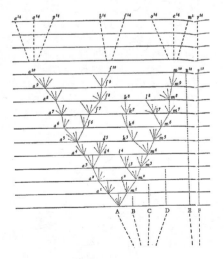

*Charles Darwin, with his
branching ancestral "tree"
diagram in 1859, was the first to
depict evolution as a complex and
multidirectional process. He took
into account the extinction of
species (B, C, and D), and
integrated humans as one branch
of evolution, rather than its
pinnacle.*

example, plants, lizards, and moray eels were all perceived as close relatives because they were…green.

Later natural historians took Aristotle's foundations in classification and sorting a step further by arranging living things in a chain of "progress" from simple (worms, insects, and such) to more complex (bears, lions, apes). Inevitably, these **chains** culminated in the "apex" of life on Earth: humans. And not just any human; early icons clearly depict a white, European male. These chains were a reflection of cultural views of the times, not factual accuracy. **Evolutionary ladders** were a modification on the chain theme. Again, these icons represented simple, early life on the lower rungs and a gradual climb to complex, "advanced" white men on the uppermost rung.

Though both the chains and ladders showed life on Earth in a sequence, they failed to depict organisms changing over time. On the contrary, these scales were frozen in time. Worms were always worms, birds always birds, and white men always white men. Since both systems depicted humans as the most advanced form of life, an organism's proximity to humans along the chain or ladder implied how "advanced" that organism was. (As an illustration of historical racism in science, there were chain and ladder icons persisting into the early 20th century that depicted certain racial groups positioned on the scale with the apes, near the bottom, and white, European males at the top.)

The Tree of Life

With the publication of Charles Darwin's *The Origin of Species* in 1859, a new metaphor for illustrating the history of life appeared. Darwin depicted the diversity of life and its change over time with a copiously branching bush or **tree.** His tree is rooted in the early life of simple cellular arrangement, and the branches show the paths of subsequent evolution. The outermost twigs and leaves depict species currently living.

What set Darwin's model apart from the chain and ladder was that life through time wasn't represented as a simple, linear sequence, but as a **complex, unpredictable** collection of twists, turns, offshoots, and dead ends. Most importantly, Darwin hoped to convey the sense that life transformed, or evolved, over time, from simple forms to complex. With this new metaphor, humans become just another twig on the tree of life (along with bacteria, orangutans, and palm trees) rather than representing

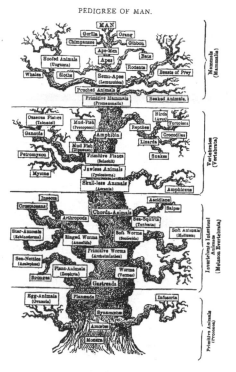

the *culmination* of all evolution to date. Darwin knew that this reassignment of human "specialness" would trouble many readers of his time. Indeed, other "trees of life" were created that put mankind squarely back at the top of the scale. One of Ernst Haeckel's trees of life not only depicted humans as supreme, but categorized them into racially superior and inferior "species." Darwin's perspective—that humans are not more "advanced" on the tree of life, just more complex at a cellular level—continues to fuel antievolutionary sentiment even today.

Creationism

> "Explanations on how the natural world changed based on myths, personal beliefs, religious values, mystical inspiration, superstition, or authority may be personally useful and socially relevant, but they are not scientific."
> —*Excerpt from the* National Science Education Standards

> "The Commission on Science Education of the American Association for the Advancement of Science is vigorously opposed to attempts by some boards of education, and other groups, to require that religious accounts of creation be taught in science classes."
> —*From the resolution passed by the American Association for the Advancement of Science Commission on Science Education, 1972*

In 1866, biologist Ernst Haeckel's "Pedigree of Man" was a more literal tree; it placed "Man" at the "top," and froze speculative linear relationships without distinguishing between extinct and living groups.

The authors and educators involved with *Life through Time* have applied the most stringent principles of science in the making of this guide. We therefore find it simplest and most persuasive to include here (following page) the exact language of the National Science Teachers Association (NSTA) on the subject of creationism, as found in "The Nature of Science and Scientific Theories."

National Science Teachers Association
"The Nature of Science and Scientific Theories"

"Creation science" is an effort to support special creationism through methods of science. Teachers are often pressured to include it or synonyms such as "intelligent design theory," "abrupt appearance theory," "initial complexity theory," or "arguments against evolution" when they teach evolution. Special creationist claims have been discredited by the available evidence. They have no power to explain the natural world and its diverse phenomena. Instead, creationists seek out supposed anomalies among many existing theories and accepted facts. Furthermore, creation science claims do not provide a basis for solving old or new problems or for acquiring new information.

Nevertheless, as noted in the *National Science Education Standards*, "Explanations on how the natural world changed based on myths, personal beliefs, religious values, mystical inspiration, superstition, or authority may be personally useful and socially relevant, but they are not scientific." Because science can only use natural explanations and not supernatural ones, science teachers should not advocate any religious view about creation, nor advocate the converse: that there is no possibility of supernatural influence in bringing about the universe as we know it.

Legal Issues

Several judicial rulings have clarified issues surrounding the teaching of evolution and the imposition of mandates that creation science be taught when evolution is taught. The First Amendment of the Constitution requires that public institutions such as schools be religiously neutral; because special creation is a specific, sectarian religious view, it cannot be advocated as "true," accurate scholarship in the public schools. When Arkansas passed a law requiring "equal time" for creationism and evolution, the law was challenged in Federal District Court. Opponents of the bill included the religious leaders of the United Methodist, Episcopalian, Roman Catholic, African Methodist Episcopal, Presbyterian, and Southern Baptist churches, and several educational organizations. After a full trial, the judge ruled that creation science did not qualify as a scientific theory (*McLean v. Arkansas Board of Education*, 529 F. Supp. 1255 (ED Ark. 1982)).

Louisiana's equal time law was challenged in court and eventually reached the Supreme Court. In *Edwards v. Aguillard* 482 U.S. 578 (1987), the court determined that creationism was inherently a religious idea and to mandate or advocate it in the public schools would be unconstitutional. Other court decisions have upheld the right of a district to require that a teacher teach evolution and not teach creation science: (*Webster v. New Lennox School District #122*, 917 F.2d 1003 (7th Cir. 1990); *Peloza v. Capistrano Unified School District*, 37 F.3d 517 (9th Cir. 1994)).

Some legislatures and policy-makers continue attempts to distort the teaching of evolution through mandates that would require teachers to teach evolution as "only a theory," or that require a textbook or lesson on evolution to be preceded by a disclaimer. Regardless of the legal status of these mandates, they are bad educational policy. Such policies have the effect of intimidating teachers, which may result in the de-emphasis or omission of evolution. The public will only be further confused about the special nature of scientific theories, and if less evolution is learned by students, science literacy itself will suffer.

The history of life is a story rich in details. Life continues to adapt and evolve over the generations, as atmosphere, climate, land masses, and bodies of water change and therefore require changes in the organisms they affect, and as plants and animals compete for survival in the changing environment. We're in the Quaternary Period of geologic time, and it isn't over yet.

What comes next? ■

Important Evolutionary Themes

This unit is extremely rich with content. Several main ideas thread through the unit and should be emphasized over the course of each session:

1. SPECIES ARE DIVERSE AND DISTINCT, BUT RELATED.

Life on Earth is hugely diverse, yet the **similarities** among organisms show that the millions of different species of plants, animals, fungi, and microorganisms that live on Earth today are related by descent from common ancestors. In this unit, even as students see a vast array of diverse organisms, the focus is on evidence of **relatedness**—similarities of characteristics like digestive tracts ("guts") or reproductive strategies. Students come to understand relatedness by thoroughly observing organisms successively through the ages, and by assembling, day by day, a classroom "tree of life."

Common structures (also known as "homologous" structures) are important evidence of evolution. Common structures share the same developmental and evolutionary origin (for example, human hand bones and corresponding bones found in bat wings). Similarities among species can be seen in external features and are even more apparent in internal organs, chemical processes, and embryonic development. (Similarities are most obvious in genetic comparisons, which is not a topic for this guide.)

2. EARTH'S PRESENT-DAY SPECIES DEVELOPED FROM EARLIER, LESS-COMPLEX SPECIES.

Over time, life has evolved NOT from "less advanced" to "more advanced," but **from simple to complex.** In *Life through Time,* this is demonstrated by studying organisms successively through the ages. The original single-celled life was very simple. Life-forms that evolved millions of years later—humans, oak trees, hippopotamuses, petunias—are extremely complex. (On the other hand, the most successful single-celled organisms—bacteria, for instance—still flourish to the present day.) The most accurate way to compare life through time is to compare the **complexity** of cellular arrangement, functions, and behaviors.

Humans are more complex, but no more "advanced," than single-celled organisms; each is very well adapted to its own particular niche. In evolutionary terms, single-celled organisms are in fact more "successful" than humans, given how long they've been around and in terms of sheer numbers.

3. The fossil record shows evidence of long periods of little or modest change, followed by relatively short bursts of great change.

The history of life is less a story of constant, gradual change than of "fits and starts," with rapid diversification and exploitation of new habitats.

4 The history of life is a story of stability as well as change.

The earliest bacterial life, many early arthropod forms, early cephalopods (*Nautilus,* for instance), and sharks were of such successful design that they've existed relatively unchanged for hundreds of millions of years.

5. The colonization of new habitats was made possible by new adaptations.

Adaptations are features or behaviors that can improve an organism's chance for survival. In order to colonize dry land, for instance, some species evolved adaptations such as "watertight" skin, a less-porous eggshell, and a water-conserving large intestine.

6. Species adapt to their environment over generations.

In discussing adaptations, it's important to point out that **species adapt, individuals do not.** Many students and adults mistakenly believe that an individual organism can adapt its features within its lifetime. In fact, adaptations occur over generations—not by choice, but by the superior reproductive success of those already born with the specific adaptations.

7. Distantly related organisms can have similar adaptive features.

Some adaptive "designs" have appeared many times throughout the history of life in distantly related species. This process is called **convergence.** Sharks, ichthyosaurs (extinct aquatic reptiles), and dolphins all have similar hydrodynamic, streamlined bodies, yet are members of very different vertebrate classes (fish, reptiles, and mammals respectively).

How does it work?
Organisms inherit particular characteristics from their parents. Humans make humans. Horses make horses. Mosses create other mosses. But each offspring displays a unique variation of these characteristics—and some even display completely novel features or behaviors. Since more offspring are produced than can possibly survive, it makes sense that those with features and behaviors best suited to survival will live to pass those traits on to their offspring, and so on. Over many, many generations, these successful characteristics may become the standard for a species—long after the original organisms are gone.

Their common body shape is the result of similar environmental pressures of making a living in an aquatic habitat.

8. By studying fossils, we can learn about organisms from the past.

Even though soft-bodied organisms leave little or no trace in the fossil record (they're usually eaten, or decompose, or dissolve away), enough fossils exist to provide evidence of adaptive structures through time.

9. The process of change and evolution occurred over vast amounts of time.

Life on Earth began over 3.5 billion years ago. It's taken millions and sometimes billions of years for new branches of organisms to evolve, and for new habitats to be colonized. In *Life through Time,* the changes in life-forms through the ages are continuously connected with geologic history.

10. Evolution builds on what already exists.

The more variety there is at a given time, the more variety there can be in the future. Organisms in the relatively short recent geologic past (the past 65 million years or so) are dramatically more diverse than they were in the long early millennia of life on Earth.

11. Evolution is also *restricted* by what already exists.

We'll never get clams with wings, or hippos with spider-like legs, because **new species build on the successful *patterns* of the species they descended from.** Each organic design is rich with evolutionary possibilities, but its paths of potential change are limited.

The rich evolutionary tapestry of the present makes perfect sense—in hindsight. We can understand how arthropods capitalized on the segmented body arrangement of annelid worms, or how mammals were able to diversify and spread into new ecosystems after the dinosaurs died out. But if we'd been dropped into any given era with no knowledge of the future, no one could have predicted these events.

12. Atmospheric and geologic changes have had a tremendous impact on life on Earth.

New species tend to evolve when members of one species are geographically split (by mountains, islands, or separating continents), or when separate species are brought together as continents collide. Whole habitats, too, can be created or destroyed with geologic and climatic change (such as when vast, warm, and shallow ocean basins covered much of North America—and again when they disappeared).

As geologic plates moved farther from or closer to the equator, weather on the continents was dramatically affected. The general patterns of weather and atmosphere, as well, affected how life evolved in each time period. (Life, in turn, can also affect the physical environment; when early plants evolved the process of photosynthesis, they made our atmosphere oxygen-rich.)

13. EXTINCTION OF SPECIES IS COMMON.

MOST OF THE SPECIES THAT HAVE LIVED ON EARTH NO LONGER EXIST. Regional **"background extinctions"** occur all the time, as small changes in climate, habitat, resources, or competitiveness prove the undoing of less-adaptive organisms. Huge **"mass extinctions"** have occurred five times in 3.5 billion years. A number of biologists think that another mass extinction is currently underway, and that up to one-half of all living species could disappear within 50 to 100 years. (This extinction would be the consequence not of astronomical phenomena or colliding continents, but of human activity, especially the destruction of plant and animal habitats.) ■

Three Billion Years plus a Few Minutes

ooze

cohesion, opulence

twinges, space, undulations

slithering, boring, hatching, germination

increase, complexity, blooming, coordination, cooperation

predation, parasitism, migration, competition, adaptation, selection

crowding, starvation, disease, populations, diversity, camouflage, mimicry

specialization, reduction, exclusion, conversion, blight, drought

pressure, defeat, decline, smoke, decimation

explosion, abuse, ignorance, teeming

contamination, residual, choking

silence, stillness

ooze.

William T. Barry

TEACHER'S OUTLINE

SESSION 1: INTRODUCING THE TREE OF LIFE

■ Getting Ready

1. Copy and prepare transparencies, cards, journal pages, and other needed sheets as described in guide.
2. Decide on overall length of adding-machine tape for **Class Time Line,** mark and cut it up into lengths for each time-travel session.
3. Post **Class Time Line** strip for Session 2, "4.5 BYA–544 MYA."
4. Mark on meterstick one million, 10 million, 100 million, and one billion years.

■ Introducing the Tree of Life

1. Tell class unit is about what life was like many millions of years ago and how it changed, or evolved, over time.
2. Hand out journal page **First Life on Earth.** Have students write their names on it.
3. Have students write what they think animals, plants, land, water, atmosphere, weather, and continents may have been like when life first appeared on Earth.
4. Ask students to share ideas. Tell them scientists think the first life-form on Earth was a single-celled organism. Write definitions of organism and cell on board.
5. Show and briefly name **Sample Tree of Life Cards** transparencies. Explain that organisms range from very "simple" (most body parts do same thing) to "complex" (different parts do different things).
6. Add definition of single-celled organism to board. Ask which organism on cards is simplest and which is most complex.
7. Ask what kinds of clues indicate that one organism is related to (descended from) another. Ask which organisms on cards might have descended from which others. Place cards in ascending order (most simple at bottom to most complex at top) according to students' guesses. **Emphasize that organisms at "top" are NOT "most advanced" or "best," just most complex.**
8. Explain that species evolve over time. Changes show in adaptations—characteristics that prove most successful for a species in its environment over many generations. Add to board definition of adaptation and statement that species evolve over many generations.
9. Show **From Worm to Insect** transparency and discuss adaptations that evolved into parts of more complex organisms.
10. Show **Tree of Life Branch—Arthropods** transparency. Say that diagram represents how scientists think these organisms are related.
11. Divide class into teams of four. Tell students their teams will go through same steps as class did, using "Tree of Life" cards at their desks. Explain procedure.
12. Provide each team an uncut **Students' Tree of Life Cards** sheet, scissors, and envelope, and let them begin.
13. When they're done, explain that team results may look different from each other. Have teams circulate to view others' work. Ask teams to share "trees of life" with class and discuss their reasoning.
14. Collect scissors and **Students' Tree of Life Cards** in envelopes.

■ Introducing Time Travel

1. Tell students they'll be time-travel scientists, visiting five time periods from Earth's past to study animal and plant life.

2. They'll observe a terrarium and an aquarium representing life from each time period.

3. Show globe or world map. Ask how old Earth is. It existed more than a billion years before first organisms appeared.

4. Ask how scientists can know about ancient organisms if organisms aren't around today. Explain that new fossil finds expand our picture of the past.

5. Tell students that first time period they'll visit includes earliest evidence of life. Period is 4.5 billion years ago to 544 million years ago. Write time period on board.

■ Introducing the Class Time Line

1. Point out posted **Class Time Line** strip you cut for next session.

2. Hold pre-marked meterstick up to time line to show how much one million, 10 million, 100 million, and one billion years takes up.

3. Say that time period class will study in next session is longest in unit.

4. Distribute **Time Travel Journal** covers and **Organism Keys.** Have students write names on journal covers and keys, then browse through Organism Keys.

5. Collect **First Life on Earth** journal pages, **Time Travel Journal** covers, and **Organism Keys.** Tell class time travel will begin next session.

6. Distribute *Life through Time* **Scavenger List for Parents** for students to take home.

SESSION 2: EARLY LIFE

■ Getting Ready

Before Day of Activity

1. Copy and prepare script, journal pages, cards, transparencies, and other needed sheets as described in guide.

2. Label containers for Organism Adaptations stations.

3. Set up **Time Travel Aquarium** and **Time Travel Terrarium** as described in guide, and decide on volcano options.

4. Set up **Life through Time** wall chart and overhead projector.

5. Set up **Tree of Life** wall chart.

On Day of Activity

1. Finish setting up terrarium volcano(es), if using.

2. Set up Organism Adaptations stations.

3. Copy station sheets and set up remaining core stations and optional additional station(s).

4. Gather students' **Time Travel Journal** covers, journal page **First Life on Earth,** and **Organism Keys.** Put out sets of journal pages labeled **Time Period #1.**

5. Set aside third copy of **Continental Drift—Time Period #1** to add to wall chart.

6. Cover aquarium and terrarium with cloths.

7. Have adding-machine-tape strip for next session ready to add to **Class Time Line.**

■ Introducing the Time Travel Journal

1. Distribute binder clips or three-ring binders for students' journals.

2. Distribute **Time Travel Journal** covers, **Organism Keys,** and **First Life on Earth** pages from previous session, and **Time Period #1** journal pages to each student.

3. Let students know they'll investigate a time period long ago, from beginning of Earth 4.5 billion years ago to 544 million years ago—a 3 billion, 956 million year stretch that represents nearly 90 percent of history of planet.

4. Tell students activities are set up at stations they will visit to record observations and answer questions in journals.

■ Preparing for Time Travel and Stations Overview

Go over station activities and materials and provide an overview to the stations as described in guide. Ask students to leave station illustrations where they are.

■ Overview Wrap-Up

1. After overview, say that at end of each day students will make predictions about next session's time period based on what they learned.

2. Say that one organism they'll study in unit is a "killer" (but quite safe to them).

■ Traveling through the Stations

1. Divide students into teams of two.

2. Assign teams to stations (up to two teams per station) and begin rotations.

3. During rotations, occasionally "fire off" terrarium volcanoes.

■ Time Travel Debrief

1. Focus students on **Life through Time** wall chart.

2. Remove backgrounds from aquarium and terrarium and tape to wall chart. Refer to them during discussion.

3. Debrief **Aquarium** and **Terrarium** stations. Ask students what they noticed about Earth during this time period, including land, water, plants, animals, weather, atmosphere, and continents.

4. Debrief the remaining stations, as described in guide, pausing after the algae reproduction debrief to fill in the **Tree of Life** wall chart.

5. Add **Continental Drift—Time Period #1** sheet to wall chart.

■ Major Evolutionary Events

1. Have students brainstorm major evolutionary changes during this period.

2. Ask students how their observations compare with their predictions.

3. Post **Major Evolutionary Events—Time Period #1** above this session's strip of **Class Time Line.**

■ Most Representative Organism of the Age

1. Ask what might be considered most representative organism of this period. Accept all answers.
2. Conduct campaign, ask for other nominations, hold election, and post results.

■ Class Time Line

1. Review major events of period, referring to **Class Time Line.** Ask how these events affect life on Earth today.
2. Attach to time line adding-machine tape for next session and write time period on board.
3. Have students write predictions in journals about changes in next period.
4. Ask students to bring in organisms for next session's adaptation stations.

SESSION 3: THE INVASION OF LAND

■ Getting Ready

One Week Ahead

Acquire *Triops* eggs, so they'll hatch before day of activity.

Before Day of Activity

1. Copy and prepare script, journal pages, transparencies, cards, and other needed sheets as described in guide.
2. Label containers for Organism Adaptations stations.
3. Update **Time Travel Aquarium** and **Time Travel Terrarium.**

On Day of Activity

1. Copy station sheets and set up remaining core stations.
2. Place organisms in containers and set up additional stations.
3. Gather students' **Time Travel Journals** and **Organism Keys.** Have **Fossils—Teacher's Answer Sheet** for this session available.
4. Put out sets of journal pages labeled **Time Period #2.**
5. Set aside third copy of **Continental Drift—Time Period #2** to add to wall chart.
6. Copy **The Age of _____** sign or write information on paper.
7. Set out transparencies **From Worm to Insect, Algae Reproduction,** and **Other Early Invertebrates** from previous sessions.
8. Cover aquarium and terrarium.
9. Have adding-machine-tape strip for next session ready to add to **Class Time Line.**

■ Preparing for Time Travel, Stations Overview, and Traveling through the Stations

1. Tell students today's time period: 544 million–410 million years ago. Write it on board or point out on **Class Time Line.** This period lasted 134 million years.
2. Distribute **Time Travel Journals,** new sets of **journal pages,** and **Organism Keys,** asking students to review what they think Earth will be like during this period. Discuss predictions. Accept all answers.

3. Provide station overviews as described in guide.
4. Divide students into teams of two, assign to stations, and begin rotations.

■ Time Travel Debrief

1. Focus students on wall chart.
2. Remove backgrounds from aquarium and terrarium and place on wall chart.
3. Debrief stations as described in guide.
4. Add **Continental Drift—Time Period #2** sheet to wall chart.

■ Major Evolutionary Events

1. Have students brainstorm major evolutionary changes during this period.
2. Ask students how their observations compare with their predictions.
3. Post **Major Evolutionary Events—Time Period #2** above this session's strip of **Class Time Line.**

■ Most Representative Organism of the Age

Tell students that they will again vote to name time period after most representative organism. Follow same campaign and election procedure as before and post results.

■ Class Time Line

1. Attach to time line adding-machine tape for next session and write time period on board.
2. Have students write predictions in journals about changes in next period.
3. Ask students to bring in organisms for next session's adaptation stations.
4. Have students place this session's **Tree of Life Organism Cards** on **Tree of Life** wall chart, or do yourself before next session.

SESSION 4: FISH AND AMPHIBIANS

■ Getting Ready

Before Day of Activity

1. Copy and prepare script, journal pages, transparencies, cards, and other needed sheets as described in guide.
2. Label containers for Organism Adaptations stations.
3. Update **Time Travel Aquarium** and **Time Travel Terrarium.**

On Day of Activity

1. Copy station sheets and set up remaining core stations.
2. Place organisms in containers and set up additional stations.
3. Gather students' **Time Travel Journals** and **Organism Keys.** Have **Fossils—Teacher's Answer Sheet** for this session available.
4. Put out sets of journal pages labeled **Time Period #3.**
5. Set aside third copy of **Continental Drift—Time Period #3** to add to wall chart.

6. Copy **The Age of** _____ sign or write information on paper.

7. Set out transparencies **Algae Reproduction** and **Moss Reproduction** from previous sessions.

8. Cover aquarium and terrarium.

9. Have adding-machine-tape strip for next session ready to add to **Class Time Line.**

■ Prepare for and Begin the Time Travel

1. Announce today's time period: 410 million–286 million years ago. Write it on board or point out on **Class Time Line.** This period lasted 124 million years.

2. Distribute **Time Travel Journals,** new sets of **journal pages,** and **Organism Keys.** Ask students to review and discuss their predictions. Accept all answers.

3. Review stations, including new ones. Divide students into teams of two, assign to stations, and begin rotations.

■ Time Travel Debrief

1. Focus students on wall chart.

2. Remove backgrounds from aquarium and terrarium and place on wall chart.

3. Debrief stations as described in guide.

4. Add **Continental Drift—Time Period #3** sheet to wall chart.

■ Major Evolutionary Events

1. Have students brainstorm major evolutionary changes during this period.

2. Ask students how their observations compare with their predictions.

3. Post **Major Evolutionary Events—Time Period #3** above this session's strip of **Class Time Line.**

■ Most Representative Organism of the Age

Conduct campaign and election as before and post results.

■ Class Time Line

1. Attach to time line adding-machine tape for next session and write time period on board.

2. Have students write predictions in journals about changes in next period.

3. Ask students to bring in organisms for next session's adaptation stations.

4. Have students place this session's **Tree of Life Organism Cards** on **Tree of Life** wall chart, or do yourself before next session.

SESSION 5: REPTILES

■ Getting Ready

Before Day of Activity

1. Copy and prepare script, journal pages, transparencies, cards, and other needed sheets as described in guide.

2. Label containers for Organism Adaptations stations.

3. Update **Time Travel Aquarium** and **Time Travel Terrarium.**

On Day of Activity

1. Copy station sheets and set up remaining core stations.

2. Place organisms in containers and set up additional stations.

3. Gather students' **Time Travel Journals** and **Organism Keys.** Have **Fossils—Teacher's Answer Sheet** for this session available.

4. Put out sets of journal pages labeled **Time Period #4.**

5. Set aside third copy of **Continental Drift—Time Period #4** to add to wall chart.

6. Copy **The Age of** _____ sign or write information on paper.

7. Set out transparencies **Algae Reproduction, Moss Reproduction, Conifer Reproduction,** and **Amphibians** from previous sessions.

8. Cover aquarium and terrarium.

9. Have adding-machine-tape strip for next session ready to add to **Class Time Line.**

■ Prepare for and Begin the Time Travel

1. Announce new time period: 286 million–65 million years ago. Write it on board or point out on **Class Time Line.** This period lasted 221 million years.

2. Distribute **Time Travel Journals,** new sets of **journal pages,** and **Organism Keys.** Ask students to review and discuss their predictions. Accept all answers.

3. Review stations, including new ones. Divide students into teams of two, assign to stations, and begin rotations.

■ Time Travel Debrief

1. Focus students on wall chart.

2. Remove backgrounds from aquarium and terrarium and place on wall chart.

3. Debrief stations as described in guide.

4. Add **Continental Drift—Time Period #4** sheet to wall chart.

■ Major Evolutionary Events

1. Have students brainstorm major evolutionary changes during this period.

2. Ask students how their observations compare with their predictions.

3. Post **Major Evolutionary Events—Time Period #4** above this session's strip of **Class Time Line.**

■ Most Representative Organism of the Age

Conduct campaign and election as before and post results.

■ Class Time Line

1. Attach to time line adding-machine tape for next session and write time period on board.

2. Have students write predictions in journals about changes in next period.

3. Ask students to bring in organisms for next session's adaptation stations.

4. Have students place this session's **Tree of Life Organism Cards** on **Tree of Life** wall chart, or do yourself before next session.

SESSION 6: BIRDS AND MAMMALS

■ Getting Ready

Before Day of Activity

1. Copy and prepare script, journal pages, transparencies, and other needed sheets as described in guide.
2. Label containers for Organism Adaptations stations.
3. Update **Time Travel Aquarium** and **Time Travel Terrarium.**

On Day of Activity

1. Copy station sheets and set up remaining core stations.
2. Place organisms in containers and set up additional stations.
3. Gather students' **Time Travel Journals** and **Organism Keys.** Have **Fossils—Teacher's Answer Sheet** for this session available.
4. Put out sets of journal pages labeled **Time Period #5.**
5. Set aside third copies of **"Guts": Bird and Mammal** and **Continental Drift—Time Period #5** sheets to add to wall chart.
6. Copy **The Age of _____** sign or write information on paper.
7. Set out transparencies **Algae Reproduction, Moss Reproduction, Conifer Reproduction, Flowering Plant Reproduction, Reptiles,** and **"Guts": Single-Celled Organism/Sponge/Jellyfish/ Earthworm** from previous sessions.
8. Cover aquarium and terrarium.

■ Prepare for and Begin the Time Travel

1. Say this is last time period they'll travel to: 65 million years ago to present time. Write it on board or point out on **Class Time Line.**
2. Ask what significant species will make first appearance on Earth in very last part of session.
3. Distribute **Time Travel Journals,** new sets of **journal pages,** and **Organism Keys.** Ask students to review and discuss their predictions. Accept all answers.
4. Review stations, including new ones. Divide students into teams of two, assign to stations, and begin rotations.

■ Time Travel Debrief

1. Focus students on wall chart.
2. Remove backgrounds from aquarium and terrarium and place on wall chart.
3. Debrief stations as described in guide.
4. Add **"Guts": Bird and Mammal** and **Continental Drift—Time Period #5** sheets to wall chart.

■ Major Evolutionary Events

1. Have students brainstorm major evolutionary changes during this period.
2. Ask students how their observations compare with their predictions.
3. Post **Major Evolutionary Events—Time Period #5** above this session's strip of **Class Time Line.**
4. Tell students that in next and final session they'll get to be inventive and express concepts they've learned throughout unit.

■ Most Representative Organism of the Age

1. Conduct campaign and election as before and post results.
2. Ask students to bring in materials for option selected for final session.

SESSION 7: REFLECTING ON LIFE THROUGH TIME

Option 1: Time Traveler Adventure Stories

Copy and prepare materials, introduce standard geologic time line and period information sheets, and set up assignment as described in guide.

Option 2: Dramatizing Life through Time

Write "starter list" of ideas on board, copy materials, decide if you will introduce standard geologic time line, and set up assignment as described in guide.

Option 3: Explosions and Extinctions

Copy and prepare materials, decide if you will introduce standard geologic time line, set up assignment, take the class through a sample question, and start students on their own questions as described in guide.

Option 4: 3-D Class Diorama

Copy and prepare materials, introduce standard geologic time line, and set up assignment as described in guide.

ASSESSMENT SUGGESTIONS

Anticipated Student Outcomes

1. Students are able to articulate, with examples, how Earth's present-day species evolved from earlier, distinct species. They can discuss how life has evolved from simplicity to complexity; for example, describing how plant reproductive structures changed from water to land and/or how animal digestive tracts evolved.

2. Students gain a deeper understanding of fossils as evidence for the existence of life in the past, and as a powerful way to learn more about those organisms.

3. Students are able to explain, with relevant examples, the concept of adaptation. They can express their understanding that species adapt to their environment or circumstances over generations, and they can describe adaptation in the context of the structures and behaviors that animals use for locomotion, feeding, digestion, and protection.

4. Students gain insight into the overarching life-science theme of disparity and unity. They're able to identify groups of animals (arthropods, invertebrates, fish, amphibians, reptiles, birds, mammals) by characteristics, and describe what makes them different and what they have in common.

5. Students are able to articulate, with examples, that major evolutionary events have occurred in Earth's past. They demonstrate awareness of the vast amount of time over which biological evolution has taken place. They are also able to express the idea that changes take place at different rates—that the fossil record shows evidence of long periods of relatively little change, followed by "short" bursts of much change. They are aware of how frequent extinction is, and that most of the species that have lived on Earth no longer exist. They increase their knowledge of the reasons for extinction, and of the tremendous impact geologic and atmospheric changes have had and continue to have on life and its evolution.

Embedded Assessment Activities

Time Travel Journal. The Time Travel Journal that students add to throughout the entire unit is an excellent and holistic assessment instrument. During the unit, students draw, identify, and label animals and plants in the Time Travel Aquarium and Terrarium. They draw the continents as geologists think they appeared during each time period. They predict and analyze adaptations of animals and plants, describe differences and similarities, describe and draw inferences from fossils, trace the path of food on intestinal diagrams, and, in general, jot down their thoughts and observations. Their summarizing assignments for the last session can be added to or written in their journals, providing a clear measure of how well they've understood the overall concept of change through time. Teacher review of the journal, during the course of the unit and upon its completion, can provide general and detailed information on student grasp of key concepts and acquisition of new learning. (Addresses outcomes 1-5)

The First Organism. In Session 1, students jot down ideas about what they think the first organism on Earth was like and what its habitat was like. These initial ideas can give the teacher insight into the knowledge and understanding students bring to the unit. (Outcomes 3, 4)

Phylogenetic Tree. In Session 1, in groups of four, students place 10 pictures of animals into a phylogenetic tree, from simplest to most complex. Teacher observation of group work and the final tree can give indications of student understanding, as well as group dynamics. (Outcomes 1, 3, 4, 5)

Culminating Unit Assignment. Session 7 offers four options for a major student assignment—each providing fertile ground for assessment. In one option, students are asked to explain a major evolutionary event in life on Earth. This can be in written form, a mural with narrative, or a skit. Another option is to imagine traveling back in time to any of 14 geologic periods and describing in detail the plant and animal life and other important features/events of that time. A third option encourages students to create a forum of their own choosing to dramatize how life has evolved over time. This can take the form of songs, poems, mnemonic phrases, skits, or any other representation they can think of that satisfies the assignment criteria. In the fourth and most elaborate option, student teams are challenged to create three-dimensional dioramas representing a period in the standard geologic time line, using information on the organisms, atmosphere and weather, and continental placement of the time.

Teacher review and comment on the assignment as it develops, and overall evaluation of the final product, can reveal how well students have integrated the conceptual underpinnings with detailed information. If, as suggested in the unit, teachers make clear to students their expectations for the assignment and the criteria on which they'll be evaluated, students will be able to keep these guidelines in mind and teachers will be able to structure their assessment on these criteria. (Outcomes 1–5)

Additional Assessment Ideas

- **Design a comic strip depicting life through time.** (Outcomes 1–5)

- **Write an advertisement or travel brochure** for a period in the geologic time line. (Outcomes 3, 4, 5)

- **The [Devonian] Gazette**
  a. Write a newspaper article about an important change in Earth's atmosphere and/or plate movements. Explain the importance of asteroids, the build-up of a new gas in the atmosphere, destruction of shallow warm seas by plate collision, etc.

  b. Write a catchy newspaper report on an evolutionary breakthrough in Earth's past that affected other life-forms in a big way. Explain the importance of the arrival of algae, dinosaur extinction, advent of flowering plants, mammals, etc. (Outcome 5)

- **Create a Creature**
  a. Design a new fictitious member of one of these groups, being sure to give it all the characteristics of the group: arthropods, invertebrates, fish, amphibians, reptiles, birds, mammals.

  b. Draw a fictitious plant or animal adapted for a specific environment (extreme heat, for example, or total darkness), and describe its adaptations and the reasons for them. Include protection, locomotion, feeding.

  c. Describe how a seed from a flowering plant might evolve if it were forced by circumstance to live in the sea, like its ancestors. (Outcomes 3, 4)

- **What My Knowledge Tells Me**
  Write a letter arguing for or against this statement: "Life has always been the way it is today, since it first started 6,000 years ago." (Outcomes 1–5)

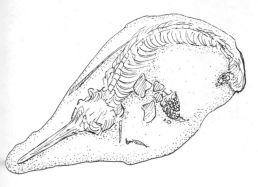

- **Unfossilize the Organism**
  Based on the illustration of this fossil, make a drawing of what you think the organism might have looked like when it was alive. (Outcomes 2, 3, 4)

- **Most Representative Creature of All Time Election**
  In an article or staged debate, argue for most representative species of all time. (Outcomes 1-5)

- **Organism Investigation**
  Describe the adaptations of an assigned animal or plant not studied in the unit and explain why you think each adaptation exists. (Outcomes 1, 3, 4)

RESOURCES & LITERATURE CONNECTIONS

Sources

Chemistry vials or tubes/Centrifuge tubes

Carolina Biological Supply
2700 York Rd.
Burlington, NC 27215-3398
(800) 334-5551
www.carolina.com

Fisher Scientific
711 Forbes Ave.
Pittsburgh, PA 15219-4785
(800) 766-7000

Nasco
901 Janesville Ave.
Fort Atkinson, WI 53538-0901
(800) 558-9595
OR
4825 Stoddard Rd.
Modesto, CA 95356-9318
(800) 558-9595

Sargent-Welch
P.O. Box 5229
Buffalo Grove, IL 60089-5229
(800) 727-4368
www.sargentwelch.com

Science Kit & Boreal Laboratories
777 East Park Dr.
P.O. Box 5003
Tonawanda, NY 14151-5003
(800) 828-7777
www.sciencekit.com

Ward's
5100 West Henrietta Rd.
P.O. Box 92912
Rochester, NY 14692-9012
(800) 962-2660
www.wardsci.com

Dry ice

Airgas Dry Ice
Penguin Brand

Distributes dry ice to thousands of grocery stores coast to coast. Call 1-877-PENGUIN for the location nearest you.

www.dryiceInfo.com

Visit for a directory of locations nationwide.

Earthworms and flatworms (Planaria)

Carolina Biological Supply
2700 York Rd.
Burlington, NC 27215-3398
(800) 334-5551
www.carolina.com

Nasco
901 Janesville Ave.
Fort Atkinson, WI 53538-0901
(800) 558-9595
OR
4825 Stoddard Rd.
Modesto, CA 95356-9318
(800) 558-9595

Sargent-Welch
P.O. Box 5229
Buffalo Grove, IL 60089-5229
(800) 727-4368
www.sargentwelch.com

Science Kit & Boreal Laboratories
777 East Park Dr.
P.O. Box 5003
Tonawanda, NY 14151-5003
(800) 828-7777
www.sciencekit.com

Ward's
5100 West Henrietta Rd.
P.O. Box 92912
Rochester, NY 14692-9012
(800) 962-2660
www.wardsci.com

Tubifex worms

Most aquarium stores carry live Tubifex worms. Look in the phone book under "Aquariums & Aquarium Supplies" or "Tropical Fish" to find stores in your area. You can also order from the following sources:

Science Kit & Boreal Laboratories
777 East Park Dr.
P.O. Box 5003
Tonawanda, NY 14151-5003
(800) 828-7777
www.sciencekit.com

Ward's
5100 West Henrietta Rd.
P.O. Box 92912
Rochester, NY 14692-9012
(800) 962-2660
www.wardsci.com

Classroom-Ready Materials Kits

Carolina Biological Supply® is the exclusive distributor of fully prepared GEMS Kits®, which contain all the materials you need for full classroom presentation of GEMS units. For more information, please visit www.carolina.com/GEMS or call (800) 227-1150.

Triops eggs

Ward's
5100 West Henrietta Rd.
P.O. Box 92912
Rochester, NY 14692-9012
(800) 962-2660
www.wardsci.com

Insect Lore
www.insectlore.com

Midnight Pass
www.midnight-pass.com

Real Cool Toys
www.realcooltoys.com

Y-Que Trading
www.yque.com

Plastic animals

Air to the Kingdom
www.airtothekingdom.org/toys.html

American Science & Surplus
www.sciplus.com

Friend's Junction
www.friendsjunction.com

Toys R Us
www.toysrus.com

- Animal Planet Dinosaur Bucket
- Animal Planet Reptile Bucket
- Flicker's Reptile Party Plastic Figurines

Related Curriculum Material

Earth History
developed by FOSS (Full Option Science System)
distributed by Delta Education
80 Northwest Blvd.
Nashua, NH 03063
(800) 258-1302

Developed at the Lawrence Hall of Science, this FOSS module for students in grades 6–8 relates to the activities in this GEMS guide. In the module, students investigate sedimentary rocks and fossils from the Grand Canyon to discover clues that reveal Earth's history. They consider the processes that created them, and compare evidence discovered in the rocks to present-day geologic processes and contemporary life-forms. Students then use these data to make inferences about past organisms, environments, and events that occurred on Earth over its history.

Ecology and Evolution
Islands of Change
by Richard Benz
NSTA Press, Arlington, VA
(2000; 224 pp.)

This book, for middle and junior high school science classes, takes big, complex ideas—how ecological pressures shape the evolutionary processes of adaptation and natural selection—and anchors them in a compelling context: the Galápagos Islands. The hands-on activities, games, and investigations will guide students to explore integrated topics from biology and earth science. Learn more about the book and view pages online at http://store.nsta.org.

How Might Life Evolve on Other Worlds?

by the SETI (Search for Extraterrestrial Intelligence)
Institute
2035 Landings Drive
Mountain View, CA 94043
(650) 961-6633

One of six teacher's guides in the Life in the Universe series, this guide for grades 5–6 helps students explore the evolution of life on Earth and search for clues to the evolution of life on an unknown planet beyond our solar system. Synthesizing what they learn, students design life-forms that could exist on that distant planet. This curriculum was supported by grants from NASA and the National Science Foundation, and involved collaboration with Lawrence Hall of Science educators. For more information, contact the SETI Institute or visit their website at www.seti.org/science/litu/litu_curriculum/Welcome.html.

Nonfiction for Students

Amazing Pop-Up 3-D Time Scape

by Stephen Biesty
Dorling Kindersley, New York, NY
(1999; 8 pp.)

This book opens out to become a poster designed to hang on a classroom or bedroom wall. It presents a visual history of the world—from the Big Bang to the present day.

Before & After
A Book of Nature Timescapes

by Jan Thornhill
National Geographic Society, Washington, D.C.
(1997; 32 pp.)

The book shows seven different habitats—tropical coral reef, wetland, savannah, meadow, temperate forest edge, rainforest, and schoolyard—each featuring a couple dozen animals. A second view of each habitat shows it in a different time period—a few seconds later, a minute later, a day later, a year later. By comparing the "before" and "after" pictures, the reader can follow how natural scenes change through time.

Before the Sun Dies
The Story of Evolution

by Roy Gallant
Macmillan, New York, NY
(1989; 190 pp.)

A complete, well-written book on the evolution of our solar system, Earth, and life.

The Cartoon History of the Universe
Volumes 1–7

by Larry Gonick
Doubleday, New York, NY
(1990; 358 pp.)

Covers the "evolution of everything" from the Big Bang to Alexander the Great. Of greatest interest are the early parts of the book: the formation of our solar system, the origins of life on Earth, and the dominant life-forms of each geologic period. Its comic book style makes it easy to read and appealing to students. Amazingly detailed, despite its broad scope.

Charles Darwin
The Life of a Revolutionary Thinker

by Dorothy Hinshaw Patent
Holiday House, New York, NY
(2001; 224 pp.)

A detailed biography of Darwin—his early life, his voyage on the *Beagle*, his life as a husband and father, and his scientific activities. The book clearly describes how, through years of careful scientific observation, Darwin developed his theory of natural selection.

The Clover and the Bee
A Book of Pollination

by Anne Dowden
HarperCollins, New York, NY
(1990; 96 pp.)

Explains the process of pollination, describing the reproductive parts of a flower and the role that insects, birds, mammals, wind, and water play in the process.

Dinosaurs and How They Lived
by Steve Parker
Dorling Kindersley, New York, NY
(1991; 64 pp.)

Describes, in text and illustrations, the physical characteristics and habits of various kinds of dinosaurs, their natural environment, and how recent discoveries have enhanced our knowledge about these prehistoric creatures.

Dinosaurs Walked Here and Other Stories Fossils Tell
by Patricia Lauber
Simon & Schuster, New York, NY
(1987; 64 pp.)

Explains what fossils are, how they are formed, how they're found and studied, and what they tell about prehistoric forms of life.

DK Nature Encyclopedia
Dorling Kindersley, New York, NY
(1998; 304 pp.)

From single-celled organisms to mammals of land and sea, this book describes the rich diversity of life on Earth. The book gives insight into how living things evolve, feed, grow, reproduce, and defend themselves. It explores each major plant and animal group, including flowering plants, birds, reptiles, insects, fish, and mammals. Thematic sections cover topics as varied as reproduction and movement, photosynthesis and communication. Also includes classification charts and a glossary of science terms.

Evolution
Life Nature Library
Time Incorporated, New York, NY
(1964; 192 pp.)

Very dated, now, but if you have this former classic on your shelf or in your library, some of the illustrations may be useful.

Evolution and the Fossil Record
by John Pojeta, Jr. and Dale A. Springer
American Geological Institute, Alexandria, VA
(2001; 36 pp.)

Provides a clear, straightforward approach to a complex topic. Includes information about the fossil record, Darwin's theory, dating the fossil record, examples of evolution, and more. Available online at www.agiweb.org/news/evolution

The Evolution Book
by Sara Stein
Workman Publishing, New York, NY
(1986; 391 pp.)

The story of 4,000 million years of life on Earth is revealed in this book through observations, experiments, projects, and investigations for children 10–14.

The Evolution of Plants and Flowers
by Barry Thomas
St Martin's Press, New York, NY
(1981; 116 pp.)

An ideal introduction to plant evolution—from single-celled algae to the invasion of land, the development of seed-bearing plants, and the complexity of flowering plants. Filled with helpful photos, drawings, and diagrams.

Eyewitness: Evolution
by Linda Gamlin
Dorling Kindersley, New York, NY
(2000; 64 pp.)

In true Eyewitness style, the book provides a good explanation of the basics of natural selection and the evolution of life on Earth through clear photographs and captions. Includes topics such as fossils, early theories, and the origin of life.

Feathered Dinosaurs

by Christopher Sloan
National Geographic Society, Washington, D.C.
(2000; 64 pp.)

Through effective and clear text, illustrations, and pictures, this book takes the reader through the fossil evidence that supports the theory that birds are the living descendants of dinosaurs. Even basic terms such as fossil and dinosaur are defined to give the reader a clear understanding. The book illustrates how scientists can learn many things from the fossil record.

Feathers

by Dorothy Hinshaw Patent
Cobblehill, New York, NY
(1992; 64 pp.)

Describes, in text and photos, birds' feathers—from structure, type, and color to various uses.

How Dinosaurs Came to Be

by Patricia Lauber; illustrated by Douglas Henderson
Simon & Schuster, New York, NY
(1996; 48 pp.)

A book about the dinosaur family tree—how dinosaurs are thought to have evolved from amphibian relatives—and the climate and geological changes that affected their evolution.

How Whales Walked Into the Sea

by Faith McNulty
Scholastic, New York, NY
(1999; 32 pp.)

A clear description of the evolution of whales—over millions of years—from terrestrial ancestors to oceanic mammals. Changes that occurred as the animals adapted to life in the sea are outlined. A border along the bottom and sides of each spread show the sea level during different eras.

Inherit the Wind

by Jerome Lawrence and Robert E. Lee
Bantam Books, New York, NY
(1960; 115 pp.)

The story of a small Tennessee town that gained national attention in 1925 when a biology schoolteacher was arrested for violating state law and teaching Darwin's theory of evolution in the classroom. See also the video listing.

Life Begins
The Detective Story of Earth's First Creatures

by John Stidworthy
Silver Burdett, Morristown, NJ
(1986; 37 pp.)

Explains what the fossilized remains of prehistoric animals have revealed about early life, from the first sea creatures to the emergence of reptiles.

Life on Earth

by David Attenborough
Little, Brown & Co., Boston, MA
(1979; 320 pp.)

Presents a history of nature, from the emergence of tiny one-celled organisms in the primeval slime more than 3,000 million years ago to primitive man. Based on the BBC television series, the book follows the sequence of highly significant events in the evolution of life on Earth.

Life on Earth
The Story of Evolution

by Steve Jenkins
Houghton Mifflin, Boston, MA
(2002; 40 pp.)

Steve Jenkins's vibrant paper collages illustrate a simple but engaging story of evolution. Geared toward younger children, but clear, concise, and effective.

Living with Dinosaurs

by Patricia Lauber; illustrated by Douglas Henderson
Bradbury Press/Macmillan, New York, NY
(1991; 46 pp.)

In prehistoric Montana 75 million years ago lived the giant reptiles and fishes of the sea; the birds and pterosaurs in the sky; the dinosaurs, tiny mammals, crocodiles, and plants of the lowlands; and the predators of dinosaur nesting grounds in the dry uplands. This book describes their relationships to each other. At the end of the book is a clear, basic description of how a fossil forms and evolves. The colorful paintings are dynamic.

The Mystery of the Mammoth Bones and How it Was Solved

by James Cross Giblin
HarperCollins, New York, NY
(1999; 128 pp.)

Tells the story of Charles Willson Peale—an artist and amateur scientist—who, with his sons in the early 19th century, excavated, reconstructed, and displayed the first complete mastodon skeleton in North America. Since, in Peale's time, dinosaur fossils had not yet been found, his work was revolutionary—it helped prove the existence of prehistoric life and raised the possibility of extinction. Overall, the book reveals how accepted knowledge has changed. Includes a geologic time scale and a short discussion of mammoth and mastodon evolution.

The Origin of Species
By Means of Natural Selection

by Charles Darwin
The Modern Library, New York, NY
(1998; 689 pp.)
(Also available in other editions, by other publishers)

Though this classic is often referred to or quoted from, not many of us have read the original. To do so, although it's dense writing, is to "get it" in Darwin's own words. For instance, of natural selection, this is what Darwin had to say:

> As many more individuals of each species are born than can possibly survive; and as, consequently, there is a frequently recurring struggle for existence, it fol-
> lows that any being, if it vary however slightly in any manner profitable to itself, under the complex and sometimes varying conditions of life, will have a better chance of surviving, and thus be naturally selected. From the strong principle of inheritance, any selected variety will tend to propagate its new and modified form.

Also available online (see listing in Internet section).

The Origin of Species
Darwin's Theory of Evolution (Words That Changed History)

by Don Nardo
Lucent Books, San Diego, CA
(2000; 112 pp.)

Careful of the title's similarity to Darwin's original! A description of the life of Darwin and how well suited he was to science due to his strengths in observation, documentation, and careful consideration. Also comments on his impact on science and society.

Roses Red, Violets Blue
Why Flowers Have Colors

by Sylvia A. Johnson;
photographs by Yuko Sato
Lerner, Minneapolis, MN
(1991; 64 pp.)

Examines the nature and function of flower colors and clearly explains their role in attracting animal pollinators to help the plants reproduce. Has great color photos as well as pictures taken in ultraviolet light to show how flowers appear to insects.

Sex in Your Garden

by Angela Overy
Fulcrum Publishing, Golden, CO
(1997; 120 pp.)

A clear and concise look at the reproductive processes of garden plants and their relationships to their propagators. Written with a flair for the humorous and a touch of the absurd, this book makes an amusing comparison between the way people and plants attract attention. It also explains why flowers are the color, size and shape they are, why they smell the way they do, and some of the extraordinary sexual adaptations plants have made.

The Sex Life of Flowers

by Bastiaan Meeuse and Sean Morris;
photographs by Oxford Scientific Films;
drawings by Michael Woods
Facts on File, New York, NY
(1984; 152 pp.)

Based on the PBS *Nature* presentation "Sexual Encounters of the Floral Kind," this book tells of the evolution of sex in plants and of the coevolution of insects/animals and plants. Also offers several examples of the "tricks" plants play on insects to ensure pollination. The many color photos and diagrams enhance the text.

The Usborne Book of Prehistoric Facts

by Annabel Craig
Usborne, London, England
(1986; 48 pp.)

Its colorful cartoon style makes it appealing to students and its quick facts make it a useful source of information.

A Walk Through Time
From Stardust to Us: The Evolution of Life on Earth

by Sidney Liebes, Elisabet Sahtouris, and Brian Swimme
John Wiley and Sons, New York, NY
(1998; 224 pp.)

Based on a traveling exhibit by the same name, this book tells the story of Earth—from the Big Bang to the first life on Earth and on through the evolution of life from primordial microbes to the recent appearance of *Homo sapiens*. A time line, which pinpoints important stages of evolution, runs throughout the book. More information and a virtual exhibit are available online at www.globalcommunity.org/wtt/.

Why Mammals Have Fur

by Dorothy Hinshaw Patent
Cobblehill, New York, NY
(1995; 26 pp.)

This book describes the many types of fur and its contribution to mammals' success. It also discusses the origins and structure of hair, the uses of fur (warmth, camouflage, defense), fur color, and human uses of animal fur.

The Young Oxford Book of the Prehistoric World

by Jill Bailey and Tony Seddon
Oxford University Press, New York, NY
(1995; 160 pp.)

Written in an easy-to-read style, this book presents a comprehensive story of the Earth—from its hot, fiery beginnings through the first life-forms to the first humans. Many excellent photos, maps, and drawings help illustrate the stages of the multi-billion-year process. Tells of fossil formation, continents drifting, life-forms of each geologic time period, and human development. Several sections have boxed features—such as early dinosaur discoverers and the asteroid theory of dinosaur extinction—which help round out the story.

Fiction for Students

An abundance of books exists on the topics of dinosaurs and evolution; we can't hope to name them all! Here's a handful of titles we thought especially good. The four books by Ruth Heller, although written for young readers, are useful for defining and explaining the reproductive strategies of plants and animals. They also illustrate how scientists group animals or plants with similar characteristics.

Animals Born Alive and Well

by Ruth Heller
Grosset & Dunlap, New York, NY
(1982; 42 pp.)
Grades Preschool-5

Almost all mammals in the world are viviparous, bearing live young rather than eggs. This is the story of both what they have in common and how they differ. See also *Chickens Aren't the Only Ones* by Ruth Heller.

The Beagle and Mr. Flycatcher
A Story of Charles Darwin

by Robert Quackenbush
Prentice-Hall, Englewood Cliffs, NJ
(1983; 40 pp.)
Grades 4-8

As an unpaid naturalist aboard the brig HMS *Beagle* on a five-year voyage around South America, Charles Darwin began to formulate his revolutionary theory of evolution. This biography includes brief descriptions of Darwin's specimen collecting (cuttlefish, shellfish fossils, sloth jawbone), scientific observation (finch beaks, an ostrich skeleton), and a subsequent eight-year study of barnacles. Although the illustrations are in a cartoon style and the writing uses a humorous approach, quite a bit of information is conveyed.

Betting on Forever

by Billy Aronson;
illustrated by John Quinn
McGraw-Hill, New York, NY
(1996; 149 pp.)
Grades 4-7

When several extinct creatures—such as a saber-toothed "tiger," a woolly mammoth, *Tyrannosaurus rex*, a giant dragonfly, and a dodo—return to Earth for a reunion, they boast about how hard life was in their day, when they had to contend with ice ages, destructive meteors, and large meat-eating creatures. However, the dodo bets that the creatures living in the current time have a harder life. The extinct animals accept the bet and set out to explore the land they thought they knew. The book offers lessons on evolution, ecology, extinction, and the delicate interdependency of all living things.

Chickens Aren't the Only Ones

by Ruth Heller
Grosset & Dunlap, New York, NY
(1981; 42 pp.)
Grades Preschool-5

Although intended for a younger audience, this book is useful for describing oviparous animals—those that lay eggs—like birds, reptiles, amphibians, fish, and insects. There are even two oviparous mammals—the spiny anteater (echidna) and the duckbill platypus, both from Australia. See also *Animals Born Alive and Well* by Ruth Heller.

Darwin and the Voyage of the Beagle

by Felicia Law;
illustrated by Judy Brook
Andre Deutsch, Great Britain
Distributed by E.P. Dutton, New York, NY
(1985; 95 pp.)
Grades 4-8

A cabin boy along on Charles Darwin's five-year voyage keeps a diary. Assisting Darwin with his collection of insect, bird, and marine-life specimens, the boy learns about their habits and habitats. On one occasion they return with 68 different species of one beetle. They collect fossils in the Andes, straddle Galápagos tortoises, discover the skeleton of *Megatherium,* and get to know the indigenous peoples. The format is oversized, with many drawings, charts, and maps.

Dinosaur Habitat

by Helen V. Griffith;
illustrated by Sonja Lamut
Avon Camelot, New York, NY
(1999; 96 pp.)
Grades 4-7

Sharing a bedroom with his pesky 8-year-old brother Ryan is bad enough for 12-year-old Nathan, but having to listen to him play with the plastic dinosaurs in a terrarium is unbearable. When Nathan tosses Ryan's fossilized egg into the terrarium, the bedroom transforms into a swampy, volcanic habitat full of giant dinosaurs and insects. During their adventure in the Jurassic environment, the boys learn a lot about dinosaurs and each other.

The Dinosaurs of Waterhouse Hawkins

by Barbara Kerley;
illustrated by Brian Selznick
Scholastic, New York, NY
(2001; 48 pp.)
Grades K-Adult

This is the story of Benjamin Waterhouse Hawkins, an artist and model builder in mid-19th-century England, and how he helped first scientists, and then the public, envision what dinosaurs actually looked like. The book gives insight to the early days of dinosaur discovery and how the fossil finds spawned a new age of scientific understanding. Extensive author's and illustrator's notes at the end of the book provide more details of the story.

The Dragon in the Cliff
A Novel Based on the Life of Mary Anning
by Sheila Cole;
illustrated by T.C. Farrow
Lothrop, Lee & Shepard, New York, NY
(1991; 212 pp.)
Grades 6-8

Written as though it were her personal journal, this is the fictionalized account of a real young woman's life in 19th-century Britain. It tells how Mary found "curiosities" (fossils) in the cliffs near Lyme Regis, England and sold them—sometimes to curious tourists, but also to scientists in the budding field of paleontology. The book clearly illustrates Mary's struggle for acceptance as a scientist in Victorian England. See also *Mary Anning: The Fossil Hunter.*

How Much Is a Million?
by David M. Schwartz;
illustrated by Steven Kellogg
Lothrop, Lee & Shepard, New York, NY
(1985; 40 pp.)
Grades K-5

With detailed, whimsical illustrations that include children, goldfish, and stars, this book leads the reader to conceptualize what at first seems inconceivable—a million, a billion, and a trillion. An adult-level note explains the calculations used. Can help students comprehend the million-year time frames discussed in the guide.

I Was a Teenage T. Rex
by Scott Ciencin;
illustrated by Mike Fredericks
(part of the Dinoverse series)
Random House, New York, NY
(2000; 192 pp.)
Grades 4-7

In this first book of the series, something goes seriously wrong with Bertram Phillips's science fair project, sending him and three other students to South

Dakota 67 million years ago—trapped in the bodies of dinosaurs. In this strange prehistoric world, the four students must learn how to use their new bodies, cope with their differences, and cooperate with each other to survive.

If You Are a Hunter of Fossils
by Byrd Baylor;
illustrated by Peter Parnall
Charles Scribner's Sons, New York, NY
(1980; 32 pp.)
Grades 3-6

The book first imagines the reader as a "hunter of fossils" finding remains throughout the United States, then portrays another hunter of fossils looking for signs of an ancient sea in the rocks of a West Texas mountain. The book goes on to poetically describe how the area may have looked millions of years ago—the warm shallow Cretaceous seas teeming with life. Reading it is an excellent way to bring fossils to life—to view them not as locked in a rock, but as ancient animals living and thriving in the sea millions of years ago.

In the Night, Still Dark
by Richard Lewis;
illustrated by Ed Young
Atheneum, New York, NY
(1988; 32 pp.)
Grades K-6

This poem emphasizes the unity of living creatures and traces the evolution of all life—in the darkest night were born the simplest creatures, then with the breaking of dawn more complex ones, and finally, metaphorically, people and day. It is based on the Hawaiian creation song, the *Kumulipo,* originally chanted at the birth of each royal child. The illustrations are striking.

The Magic School Bus
In the Time of the Dinosaurs
by Joanna Cole;
illustrated by Bruce Degen
Scholastic, New York, NY
(1994; 44 pp.)
Grades 3-6

Just as the students are ready to show off their dinosaur projects at the school's Visitors Day, Ms. Frizzle announces they'll be going to a dinosaur dig instead. After the class learns how paleontologists uncover fossils, Ms. Frizzle turns the bus into a time machine and they're transported back through time. They visit the many different time periods in which dinosaurs lived. Diagrams help the reader keep track of the time period the class is visiting. The book is packed with information about different types of dinosaurs, plants of the time, shifting continents, and definitions of key terms.

Mary Anning
The Fossil Hunter
by Dennis B. Fradin;
illustrated by Tom Newsom
Silver Press, Parsippany, NJ
(1997; 30 pp.)
Grades 4–8

A very complete and scientifically accurate biography of Mary Anning who, as a young girl in the early 1800s, discovered the complete skeleton of an ichthyosaur on a beach near Lyme Regis, England. The book provides a synopsis of her life as a fossil hunter and an overview of the times in which she lived (the field of paleontology was young and the term "dinosaur" was just being coined). The final pages of the book contain photographs of some of the fossils Mary found. See also *The Dragon in the Cliff* by Sheila Cole.

Plants That Never Ever Bloom
by Ruth Heller
Grosset & Dunlap, New York, NY
(1984; 42 pp.)
Grades Preschool–5

This book is all about the plants that don't have flowers and propagate by seeds, spores, and cones. The plants—known as gymnosperms—include lichens, seaweed, and mosses. See also *The Reason for a Flower* by Ruth Heller.

The Reason for a Flower
by Ruth Heller
Grosset & Dunlap, New York, NY
(1983; 42 pp.)
Grades Preschool–5

In this book the purpose for a flower is given and diverse forms of flowering plants are shown. It focuses on the function of pollen and seeds, the different methods of seed dispersal, and gives some examples of atypical flowers. The term angiosperm is introduced to mean flowering plants. See also *Plants That Never Ever Bloom* by Ruth Heller.

Right Here on This Spot
by Sharon Hart Addy;
illustrated by John Clapp
Houghton Mifflin, New York, NY
(1999; 32 pp.)
Grades K–3

While digging a ditch in his cabbage field, Grandpa uncovers some items (a button, an arrowhead, bones) that provide clues about what events took place in the area in the past. Although meant for a younger audience, the book helps show how an area can change over time, and that artifacts teach us about the past.

Wildside
by Steven Gould
Tor Books, New York, NY
(1996; 316 pp.)
Grades 6–9

When Charlie, who just graduated from high school, inherits a ranch from his uncle, he is amazed to discover that a secret gate in the back of the barn leads to a "pristine, uninhabited parallel Texas stocked with extinct megafauna" but without humans. Charlie and some friends decide to exploit the wild side of his ranch by selling extinct passenger pigeons to zoos and mining for gold, but they attract the unwanted attention of government intelligence agents. An exciting adventure sure to appeal to young adults. (Note that this Steve Gould is not Stephen Jay.)

Your Mother Was A Neanderthal
by Jon Scieszka;
illustrated by Lane Smith
(part of the Time Warp Trio series)
Viking, New York, NY
(1993; 78 pp.)
Grades 4–7

In this book from the series by a popular author/illustrator team, the Time Warp Trio (three time-traveling boys) travel back to the Stone Age. Their adventures include hiding from a saber-toothed cat and running from a woolly mammoth. Filled with humor and light-hearted suspense.

Nonfiction for the Teacher

At the Water's Edge
Fish with Fingers, Whales with Legs, and How Life Came Ashore but Then Went Back to Sea
by Carl Zimmer
Touchstone, New York, NY
(1999; 304 pp.)

Describes recent fossil discoveries, outlines an evolutionary chronology, and gives insights into macroevolution.

The Boilerplate Rhino
Nature in the Eye of the Beholder
by David Quammen
Touchstone Books/Simon & Schuster, New York, NY
(2001; 288 pp.)

A collection of columns originally written for *Outside* magazine. The unifying theme is how humans regard and react to the natural world.

Evolution
The Triumph of an Idea
by Carl Zimmer
HarperCollins, New York, NY
(2001; 320 pp.)

This is the companion book to the *Evolution* series shown on PBS (see listing in Videos section). It traces the development of the theory of evolution, provides an overview of most of the topics needed to understand evolution, and summarizes changing scientific views.

Full House
The Spread of Excellence from Plato to Darwin
by Stephen Jay Gould
Random House, New York, NY
(1997; 256 pp.)

Gould argues persuasively against the concept of a simple evolutionary march toward ever-increasing complexity. New (and successful) species, he argues, are just as likely to be extremely simple as more complex.

Insects and Flowers
The Biology of a Partnership
by Friedrich G. Barth;
translated by M. A. Biederman-Thorson
Princeton University Press, Princeton, NJ
(1985; 297 pp.)

Explains how flowers exist not for our pleasure but as a triumph of coevolution with insects. A great introduction to the relationship between flowers and insects, this book shows how the colors, shapes, and scents of flowers ensure pollination in ingenious and fantastic ways. Has two great chapters on the "tricks" orchids play on insects.

Song of the Dodo
Island Biogeography in an Age of Extinctions
by David Quammen
Touchstone Books/Simon & Schuster, New York, NY
(1997; 704 pp.)

By applying the lessons of island biogeography (the study of the distribution of species on islands and island-like patches of landscape) to modern ecosystem decay, this book offers insight into the origin and extinction of species, the relationship of humans to nature, and the future of the world.

Summer for the Gods
The Scopes Trial and America's Continuing Debate over Science and Religion
by Edward J. Larson
Basic Books, New York, NY
(1997; 318 pp.)

Unlike the play and movie *Inherit the Wind* (see Videos section), this book de-romanticizes the Scopes "monkey trial" of 1925, describing the larger societal context that ignited the whole event. Fascinating, and winner of the Pulitzer Prize for history.

Teaching about Evolution and the Nature of Science
by the National Academy of Sciences
National Academy Press, Washington, D.C.
(1998; 150 pp.)

Provides a well-structured framework for understanding and teaching evolution. Written for teachers, parents, and community leaders as well as scientists and educators, this book describes how evolution reveals both the great diversity and the similarity among the Earth's organisms; it explores how scientists approach the question of evolution; and it illustrates the nature of science as a way of knowing about the natural world. In addition, the book provides answers to frequently asked questions to help readers understand many of the issues and misconceptions about evolution. Includes sample activities for teaching about evolution and the nature of science. Also available online at www.nap.edu/readingroom/books/evolution98/.

Videos

Evolution series

A PBS television series co-produced by WGBH Boston and Clear Blue Sky Productions. The series examines evolutionary science and the profound effect it has had on society and culture. From the genius and torment of Charles Darwin to the vast changes that spawned the tree of life, from the role of mass extinctions in the survival of species to the power of sex to drive evolutionary change, *Evolution* is fascinating and far-reaching in scope. The series also explores the emergence of consciousness, the success of humans, and the perceived conflict between science and religion in understanding human life. An excellent Teacher's Guide is available to accompany the videos. See also listing in Internet section.

Inherit the Wind

This is the title of both a short play by Jerome Lawrence and Robert E. Lee and a 1960 movie starring Spencer Tracy. Both forms have been produced in a variety of formats, and can often be found at local libraries. Based on the so-called Scopes "monkey trial," it's the story of a small Tennessee town that gained national attention in 1925 when a biology schoolteacher was arrested for violating state law and teaching Darwin's theory of evolution in the classroom. See also the book listing in Nonfiction for Students.

The Shape of Life series

A PBS television series produced by Sea Studios Foundation for National Geographic Television & Film. This eight-part series answers questions about the origins of life on Earth. It tells the story of the dramatic rise of the animal kingdom and helps make sense of the diversity of animal life that exists today. See also listing in Internet section.

Allegro non Tropo
Directed by Bruno Bozzetto

Much of this Italian classic isn't for young people, but the animated dinosaur migration and extinction sequence, set to Ravel's *Bolero,* is beautiful and moving.

Fantasia (the original, 1940)
Disney

The animated masterpiece from Walt Disney, with a dramatic and unforgettable segment on the collapse of the dinosaurs' reign.

CD-ROM

Explorations Through Time
CSTA Publications
3800 Watt Ave., Suite 100
Sacramento, CA 95821
(916) 979-7004

Includes 7 modules focusing on the major science concepts of evolution. Based on an interactive web experience developed by the UC Museum of Paleontology, the CD is for teachers and students without an Internet connection. Available exclusively through CSTA (California Science Teachers Association). See the Internet sites section for the online version.

Internet sites

Teaching about Evolution and the Nature of Science
www.nap.edu/readingroom/books/evolution98/

A virtual voyage of the *Beagle* activity
www.biology.com/visitors/ae/voyage/introduction.html

The Bungee Jumpin' Cows
website and song samples for *Life through Time*
www.moo-boing.com

Online versions of Darwin's books
www.literature.org/authors/darwi-charles

Dating Rocks and Fossils
www.museum.vic.gov.au/dinosaurs/fossdate.stm

The Dawn of Animal Life—an online exhibit from Miller Museum of Geology
geol.queensu.ca/museum/exhibits/dawnex.html

Website for *Evolution* television series
www.pbs.org/evolution

Explorations Through Time—interactive modules on history of life on Earth
www.ucmp.berkeley.edu/education/explotime.html

Getting Into the Fossil Record
www.ucmp.berkeley.edu/education/explorations/tours/fossil/

The History of Life as Revealed by the Fossil Record
www.ultranet.com/~jkimball/BiologyPages/G/GeoEras.html

First Dinosaur Fossil Discoveries
www.enchantedlearning.com/subjects/dinosaurs/dinofossils/First.shtml

Information on geological time scales and the organisms of each time
www.ucmp.berkeley.edu/exhibits/geology.html

Introduction to Relative Time Scale and the Age of the Earth
pubs.usgs.gov/gip/geotime/contents.html

The University of California Berkeley Museum of Paleontology's "History of Life" site
www.ucmp.berkeley.edu/historyoflife/histoflife.html

Links to Sites on Geologic Time, Dinosaurs, Fossils, Evolution, and Dating Techniques
http://personal.cmich.edu/~franc1m/geoltime.htm

National Center for Science Education: Defending the Teaching of Evolution in the Public Schools
www.natcenscied.org/

Links to many sites related to paleontology
http://members.aol.com/fostrak/kpaleo.htm

Maps showing break-up of Pangaea
http://pubs.usgs.gov/publications/text/historical.html

Animated Views of Continental Drift
http://www.ucmp.berkeley.edu/geology/tectonics.html

or

http://www.ucmp.berkeley.edu/geology/tecall1_4.mov

NSTA's Resources for Teaching Evolution
http://pubs.nsta.org/galapagos/resources/page1.html

The National Museum of Natural History's Paleobiology site
www.nmnh.si.edu/paleo

Website for *The Shape of Life* television series
www.pbs.org/kcet/shapeoflife/

Tree of Life Web Project
http://tolweb.org/tree/phylogeny.html

A Walk Through Time: From Stardust to Us
www.globalcommunity.org/wtt/walk_online.html

REVIEWERS

We warmly thank the following educators, who reviewed, tested, or coordinated the trial tests for *Life through Time* in manuscript or draft form. Their critical comments and recommendations, based on classroom presentation of these activities nationwide, contributed significantly to this GEMS publication. (The participation of these educators in the review process does not necessarily imply endorsement of the GEMS program or responsibility for statements or views expressed.) Their role is an invaluable one; feedback is carefully recorded and integrated as appropriate into the publications. WE THANK THEM ALL! ■

CALIFORNIA

Park Middle School, Antioch
Colleen Leong
★Debi Molina
Becky Ruth
Sandy Weigand

Creekside Middle School, Castro Valley
Mary Cummins
Tarri King
Cathleen Lloyd
★Scott Malfatti

Al Schilling School, Newark
★Katharine Keleher

Musick Elementary, Newark
Milene Rawlinson
Jan Schmitt

St. Paschal School, Oakland
Johnna Grell
★Sally Simpson
LaShonda Taylor
John Urbanski

Walt Disney School, San Ramon
★Judy Adler
★Barbara Bureker
Linda Debus
Kris Macias
Ed Medford

Springstown School, Vallejo
★Johanas DeBruin

ILLINOIS

KIDS, Loves Park
★Brenda Cox

Locust School, Marengo
Sue Ellen Wildens

MAINE

C.K. Burns School, Saco
Andrea Cole
Carla Cote
Carrie Lamothe
Janice Loughlin

MMSA Office, Scarborough
★Andy Vail

MICHIGAN

Marysville High School, Marysville
Stephanie McCool
Cherie Rockwell
Todd Shivers
Ken Vineyard

Washington Elementary, Marysville
★Sheryl Fraley

NEW YORK

Bronx Outreach, Flushing
Cynthia Loran

Queens Academy, Flushing
Chris Brown
Lynn Plunkett
★Steve Zbaida

Lower Manhattan Outreach, New York
Linda Brown

TEXAS

Encino Park School, San Antonio
Mary Andersen
Brande Brice
★Sandy Geisbush
Jaan Talsma

★ Trial test coordinators

Get Connected – Free!

Get the *GEMS Network News*,

our free educational newsletter filled with...

- **updates** on GEMS activities and publications
- **suggestions** from GEMS enthusiasts around the country
- **strategies** to help you and your students succeed
- **information** about workshops and leadership training
- **announcements** of new publications and resources

Be part of a growing national network of people who are committed to activity-based math and science education. Stay connected with the **GEMS N**etwork News. *If you don't already receive the* **Network News,** *simply return the attached postage-paid card.*

For more information about GEMS call (510) 642-7771, or write to us at GEMS, Lawrence Hall of Science, University of California, Berkeley, CA 94720-5200, or gems@uclink4.berkeley.edu.

Please visit our web site at www.lhsgems.org.

GEMS activities are effective and easy to use. They engage students in cooperative, hands-on, minds-on math and science explorations, while introducing key principles and concepts.

More than 70 GEMS Teacher's Guides and Handbooks have been developed at the Lawrence Hall of Science — the public science center at the University of California at Berkeley — and tested in thousands of classrooms nationwide. There are many more to come — along with local GEMS Workshops and GEMS Centers and Network Sites springing up across the nation to provide support, training, and resources for you and your colleagues!

Yes!

Sign me up for a free subscription to the

GEMS Network News

filled with ideas, information, and strategies that lead to Great Explorations in Math and Science!

Name_____

Address_____

City_____ State_____ Zip_____

How did you find out about GEMS? (Check all that apply.)
❑ word of mouth ❑ conference ❑ ad ❑ workshop ❑ other: _____
❑ In addition to the *GEMS Network News*, please send me a free catalog of GEMS materials.

GEMS
Lawrence Hall of Science
University of California
Berkeley, CA 94720-5200
(510) 642-7771

Ideas ◀

Suggestions ◀

Resources ◀

that lead to Great Explorations
in Math and Science!

LHS GEMS

01 LAWRENCE HALL OF SCIENCE # 5200

1571-25775-62-X

BUSINESS REPLY MAIL
FIRST-CLASS MAIL   PERMIT NO 7   BERKELEY CA

POSTAGE WILL BE PAID BY ADDRESSEE

UNIVERSITY OF CALIFORNIA BERKELEY
GEMS
LAWRENCE HALL OF SCIENCE
PO BOX 16000
BERKELEY CA 94701-9700

NO POSTAGE
NECESSARY
IF MAILED
IN THE
UNITED STATES

Get Connected!
www.lhsgems.org

Gusano

© 1993 Mel McMurrin & Kevin Beals
from Rock Candy by The Bungee Jumpin' Cows
www.moo-boing.com

(I Am A Worm)

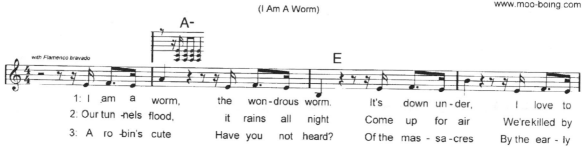

with Flamenco bravado

1: I am a worm, the won-drous worm. It's down un-der, I love to
2: Our tun-nels flood, it rains all night Come up for air We're killed by
3: A ro-bin's cute Have you not heard? Of the mas-sa-cres By the ear-ly

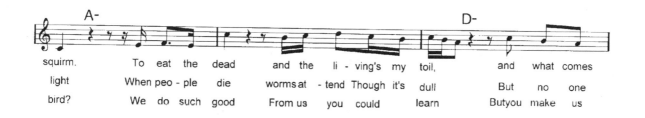

squirm. To eat the dead and the li-ving's my toil, and what comes
light When peo-ple die worms at-tend Though it's dull But no one
bird? We do such good From us you could learn But you make us

out makes mag - ni-fi-cent soil. Gu-sa no- no- no- no- no- no- no- no- no-
comes to a worms' fu - ne-ral
fish bait in re-turn

no - no - no - no - no - no. Sí! Gu-sa- no- no - no no - no- no- no- no- no-

no- no- no - no- no- no. Sí! I aer-ate the earth as I tun-nel and squirm I am

proud to be called a worm

Arachnidae

Medium tempo funk

I think I hear the door - bell   I guess it's time to eat   My
2. Webs can be tri -an-gu lar -   fun - nel - shaped or   round
3. When it's time to mate   its a dan gerous dat -ing game   Some

web it is a-jig - glin' from some in -sects' ti -ny feet   I'll
Some have stick - y trip lines,   some dig down in the ground
spiders eat each o -ther   things that wiggle are all the same   The

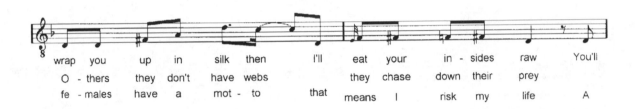

wrap you up in silk then   I'll eat your in -sides raw   You'll
O -thers they don't have webs   they chase down their prey
fe -males have a mot - to that   means I risk my life   A

be my in -sect smooth -ie,   suck your guts like with a straw   It's
I'm so glad to meet you but I   guess it's not your day   It's
hus -band's place is in -side   the sto -mach of his wife   It's

kil -ler tag and I'm it,   don't you want to play   I just want to wish you an a-

rach - ni -dae

Dancing With The Insects

(Bailando Con Los Insectos)

it did-n't stop until I was the size of a bug (Go to verse 2)
said, "hey man aren't you the bugs that I used to squash? I was
so I squish no-bo - dy I do it care - ful - ly

danc-ing with the in-sects Bai - lan - do con los in - sec - tos I was Ven - ga ju-gar.

¡Vamos a bai - lar y gozar!

Shark's in the House

© 1994 Kevin Beals & Mel McMurrin
from Rockin' the Foundations of Science by
The Bungee Jumpin' Cows www.moo-boing.com

Rap: slow, spooky, hip-hop groove

There's miles of ocean underneath me
What's messin' with my mind is that I just can't see
What's down below lookin' up at me
Thinkin' of what a tasty meal I could be
From out of the depths torpedo slippin' through the blue
Them beady black eyes have a great view
Of me with my little legs adanglin', don't want a manglin'
From you an unscheduled rendezvous
How do you do? You cordon bleu
No way - uh, uh - I'm outa here
And let me make one thing perfectly clear
You see when it comes to sharks and my backside
You call me chicken, I'll say "yeah, Kentucky Fried"

Watch out! Watch out! Shark's in the house! (4x)

You been cruisin' the salty blue H$_2$O
As predator king since 400 million years ago you know
You make no bones about it you ain't got none
You're just a cartilage creature causin' 911's,
You're no fun at a party, cause you feed in that frenzied way
Can't take you anywhere, just leave me outa your buffet
And don't smile at me with those steak knives
Your conveyor belt teeth they ruin skin dives
You know what I'm sayin'? I wanna stay alive, but you know right
where I'm at
Your sensory could locate me in nothin' flat
Look, when the scene is marine, I'm a lousy source of protein
So save your guillotine, I'm just lean cuisine

What kind of a shark are you? I wanna know before I die
Some sharks can terrify, but some are plain shy
Carcharodon megalodon could eat a car they say
So what are you a hammerhead, Great White or San Jose?
You must be 50 feet long and weigh 20 tons
A whale of a shark comin' at me I know I'm done for
Say what, Mr. Big stuff
You eat plankton and that's enough?
I was scared for nothin' guess it's true what they say
Sharks they prefer much more tasty prey—they're gourmet
Now it's true there's a few that might take a bite
But for most a human sandwich spoils their appetite
You sharks are all right, I guess you have your place
So live long and prosper, but stay outa my face!
Hey you with the dorsal, I ain't your morsel
You can just take your slicin' and dicin' bootay someplace else
Uh-oh, here comes el fin ....

Bubba
(The Frog)

© 1993 Kevin Beals & Pete Madsen
from Rockin' The Foundations of Science by
The Bungee Jumpin' Cows. www.moo-boing.com

Slow Swampy Shuffle

Let me tell you 'bout my frog-gie friend He lives in the wa-ter and he lives on land He's an am-phi-bi-an, he lives in-be-tween He's got a big mouth and his skin is green His name is Bub ba, he's a big fat frog Been eat in' bugs all day here on this log The sky is dark, it's get ting late He'd better croak loud, if he wants a date

We Don't Wanna Go

(The Dinosaur Graduation Song)

One Of These Days We're Gonna Rule The World
(The Cockroach)

© 1993 Peter Madsen, Kevin Beals & Mel McMurrin
from Rockin' the Foundations of Science by
The Bungee Jumping Cows. www.moo-boing.com

Antenna banging moderato

Life used to be soft, till the shells gave them might The seas were ruled by the

tri - lo - bites i - mag - ine be - ing grabbed by some - thing sev - en feet long Sea

scor - pi - ons were nas - ty but now they're gone De - vo - ni - an was ruled by fish Now most

of them are just a si - de di - sh So what if the sharks are still around ?
Hey tough guy,

come on up on the ground - HA! One of these da - ys we're gon - na rule the wo - rld

we al - read - y do One of these da - ys we're gon - na rule the wo - rld

we al - read - y do

Family Tree

© 1993 Kevin Beals & Mel McMurrin
from Rockin' the Foundations of Science by
The Bungee Jumpin' Cows. www.moo-boing.com

Medium tempo African Juju

A lit-le cell ap-peared a-mong the che-mi-cals one day the ma-ny celleds soon ate it up and then they swam a-way 'till a ma-ny-celled with mus-cle came and ate them as hors d'oeuvres this jel-ly fish has quite a mouth and had a lot of nerves oh yes this is my fa-mi-ly tree oh yes it's e-vo-lu-tion-a-ry oh yes we are fam-mi-ly oh yes I've got all my cou-sins and me

1. MONKEY PUZZLE TREE
2. *ARCHAEOPTERYX*
3. *SMILODON*
4. GLACIER
5. CONDOR
6. SEQUOIA
7. WOOLLY MAMMOTH
8. *EQUUS*
9. *PLIOHIPPUS*
10. *HYRACOTHERIUM*
11. *DICROCERUS*
12. CATTAILS
13. WATER LILIES
14. PLACODERM
15. SALMON
16. *DOLICHOSOMA*
17. EEL
18. SEASTARS
19. TRILOBITES

20. *WIWAXIA*
21. *HALLUCIGENIA*
22. CRINOIDS
23. SPONGES
24. AMMONITE
25. EURYPTERID
26. JELLY
27. *PROTOCARIS*
28. STROMATOLITES
29. ICHTHYOSTEGA
30. GIANT COCKROACH
31. *MEGANEURA*
32. MILLIPEDE
33. HORSETAILS
34. BAGACERATOPS
35. CYCADS
36. PTEROSAUR
37. *BRACHIOSAURUS*
38. GIANT CLUB MOSS